How to Outthink, Outmaneuver, and Outperform Your Competitors

Lessons from the Masters of Strategy

How to Outthink, Outmaneuver, and Outperform Your Competitors

Lessons from the Masters of Strategy

Norton Paley

CRC Press
Taylor & Francis Group
Boca Raton   London   New York

CRC Press is an imprint of the
Taylor & Francis Group, an **informa** business

A PRODUCTIVITY PRESS BOOK

CRC Press
Taylor & Francis Group
6000 Broken Sound Parkway NW, Suite 300
Boca Raton, FL 33487-2742

Printed on acid-free paper
Version Date: 20130321

International Standard Book Number-13: 978-1-4665-6540-1 (Hardback)

Library of Congress Cataloging-in-Publication Data

Paley, Norton.
    How to outthink, outmaneuver, and outperform your competitors : lessons from the masters of strategy / Norton Paley.
    pages cm
    Includes bibliographical references and index.
    ISBN 978-1-4665-6540-1 (hardcover : alk. paper)
    1. Strategic planning. 2. Competition. I. Title.

HD30.28.P285 2013
658.4'012--dc23   2013002111

Visit the Taylor & Francis Web site at
http://www.taylorandfrancis.com

and the CRC Press Web site at
http://www.crcpress.com

To the next generation:

Isaiah, Natalia, Zeke

Contents

Introduction

The following actual headlines appeared in the business press:

Google Targets Microsoft

The Internet company builds a new computer operating system and attacks Microsoft in virtually all its businesses.

Strategy response: Microsoft fights back by offering a free, Web-based version of its Office™ software.

Panasonic Reaches Wide with Appliances for Emerging Markets

The Japanese company wants a bigger share of sales in developing markets, but knows that its $1,200 large-screen plasma-display TV sets and $3,000 nanotechnology refrigerators are beyond the reach of low-wage families. The company's new line of offerings includes TVs for $50, air conditioners for $100, and washing machines in the $200 range.

Strategy response: Relying on years of experience squeezing costs and working on paper-thin margins, local manufacturers respond with similar products at bargain prices.

TiVo Wants to Be the Google of Television

The upstart company popularized a technology that makes it routine for viewers to save TV shows on a hard drive so they can watch them later and fast-forward through the commercials.

Now it threatens to disrupt the industry and competitors by trying to remake itself as the Google of television.

Strategy response: Cable companies counterattack by rolling out their own digital video recorders and taking subscribers away from TiVo.

AT&T Looks to Turn Mobile Access into a Big Business

AT&T assembled a team to operate like a start-up. Mission: Come up with innovative ways for people to use its wireless network. Strategy: Go beyond cell phones and hook up all manner of electronics to the Internet—from digital cameras and navigation devices to parking meters.

Strategy response: Verizon and other rivals pursue similar avenues of growth by selling inexpensive netbooks and signing up customers to their networks.

What are the commonalities among these company examples?

Managers at all the firms strained to reach the hard numbers, as measured by such metrics as sales, market share, and profitability. They all persisted in ongoing efforts to (1) engage customers and maintain secure relationships, (2) protect their customers from rivals' take-over strategies, and (3) prevent competitors from interfering with their overall growth plans. Thus, the battle lines were drawn as each manager attempted to outthink, outmaneuver, and outperform the others through innovative strategies and tactics.

The essential point in these examples: The mind of one manager is pitted against the mind of the other. As each individual faces off against his or her rival through actions and counteractions, success ultimately is determined by the skillful application of competitive strategies.

The central purpose of this book, therefore, is to equip you with a firm grasp of the roots of strategy, to understand how strategies are developed, and to apply them to a range of volatile competitive conditions.

ROOTS OF STRATEGY

The origins of the word *strategy* come from the ancient Greeks and are derived from the word *strategia* (or *strategos*), meaning to lead an army or generalship. Commanders over the centuries have relied on military strategy to conquer territory and gain power. That meant imposing their will on others and maximizing the impact of their economic and human resources to achieve their goals.

The generals faced formidable challenges as they crafted plans to outmaneuver competing forces, gain territory and power, and conserve resources, while expanding their influence. To impose their wills on others and achieve their objectives, they had to distract and unbalance their opponents physically and psychologically. Faced with resistance, those leaders were forced to maximize the effectiveness of their resources to achieve their goals.

Although the terminology varies, these challenges are not much different from those of other human endeavors—whether business, politics, or athletics. Most confrontations involve a defense protecting the ground and an offense trying to overtake that ground—or in business terminology, securing a market, as well as influencing a group's behavior.

Although the destructive aspects of war are not present in business, there is a reasonable parallel when one considers the bankruptcy of organizations, including once-mighty global leaders; the vast layoffs of thousands of personnel; the closing of physical plants, with the devastating economic impact and societal disruption that create demoralizing misery among large groups of individuals. In many instances those powerful shocks result in decimated regions, cities, and local communities.

Many business scholars, executives, and line managers readily accept the military–business connection and find practical wisdom in studying the chronicles of military conflicts that span 2,500 years of recorded history. By examining the strategic and human elements of clashes, they gain valuable insights that provide an additional dimension to the study of business.

In particular, the lessons gleaned from military history and strategy can be indispensable in the pragmatic, everyday management of people and resources, especially when applied to competitive issues. It is through the long lens of time and space that this additional perspective can fortify your judgment in such areas as leadership and employee behavior.

The further aim of this book, therefore, is to tap the universal logic and historic lessons of strategy to uncover solutions for today's stubborn competitive problems. Doing so overrides the narrower pathway of focusing only on current business events or job experience, where viewpoints tend to limit the range of opportunities and reduce judgment to a relatively short-term, constricted outlook.

Consequently, the military viewpoint provides foundation principles that can strengthen your understanding of strategy, whether you operate in a multinational firm maneuvering for position in a global arena, or a regional business fighting an everyday battle for survival. It is in this framework, then, that the rules, concepts, and strategies are presented in the following chapters.

The classic works of the masters of strategy are woven throughout: Sun Tzu's *The Art of War* and Carl von Clausewitz's *On War,* as well as the writings of Jomini, Mao Tse-tung, Machiavelli, and twentieth-century historian B. H. Liddell Hart. Their collective works provide longevity, general acceptance, and authority to the workings of strategy, which are attributed to two main factors: First, the underlying patterns of human nature have not changed significantly throughout history. Second, notwithstanding geographic, political, and technological changes, the pragmatic logic supporting strategy retains its value today as it has since the writings of Sun Tzu 2,500 years ago.

Some fields, particularly those in science, medicine, technology, and economics, are continually updated in response to breakthrough discoveries. In contrast, many of the historic treatises in the humanities still remain as noteworthy and valuable as the day they were written.

It is the study of warfare, however, that fits a special category. Fighting, combat, feud, rivalry, competition, and power struggles give a human face to conflict. Thus, by focusing on the distinguishing qualities of human behavior, leadership, and organizational culture, you can benefit from a broader perspective that leads to a better understanding of your problems. In turn, you hone the skills for applying appropriate strategies to outmaneuver competitive obstacles.

APPLYING THE LESSONS OF MILITARY STRATEGY TO BUSINESS PRACTICE

Specifically, then, how does the study of military strategy link to business, so that the time-tested practices, guidelines, rules—and the overall body of knowledge—can be transferred and applied?

We begin with the everyday language of business. It is not uncommon to read in the business press, or hear at seminars and speeches, the war-like vocabulary borrowed from the military, with such phrases as attacking a competitor, developing a strong position, defending a market, strengthening logistics, deploying personnel, launching a campaign, developing a strategy, utilizing tactics, coping with price wars, doing battle with … and other familiar comments.

In addition, there are the more indirect references that connect the military with business, such as holding reserves to exploit a market advantage, developing an intelligence network to track a competitor's actions, avoiding direct confrontation with the market leader, bypassing a market because of high entry barriers, reorganizing the marketing and sales effort to strengthen a market position, or employing a new technology to create a competitive advantage over a weaker rival.

Yet in all of the foregoing discussion, fighting may appear to be the object of warfare. And aggressive business competition may appear to be the endgame. Just the opposite is true, however, as noted by the following convincing statements:

> The object of war is a better state of peace.
>
> **B. H. Liddell Hart**

> The object of business is to create a customer.
>
> **Peter Drucker**

> For to win one hundred victories in one hundred battles is not the acme of skill. To subdue the enemy without fighting is the acme of skill.
>
> **Sun Tzu**

Given the origins of *strategy*, now we can define the term as it applies to business. Although the term is freely applied throughout an enterprise, various definitions of business strategy exist. The following meaning is used in this book: Business strategy is the art of coordinating the means (money, human resources, and materials) to achieve the ends (profit, customer satisfaction, and company growth) as defined by company policies and objectives.

In more pragmatic terms, strategy consists of *actions* to achieve *objectives* at three distinct levels:

First, *corporate strategy*. At this level, strategy is developed at the top echelons of the company. The aim here is to deploy resources through a series of actions that would fulfill the vision and objectives as expressed in a long-term strategic business plan.

Second, *mid-level strategy*. At this juncture, strategy operates at the department or product line level. Its time frame is more precise than corporate strategy. Typically, these strategies cover actions in a three- to five-year period and focus on fulfilling specific objectives. Also at this level, strategy embodies two zones of activities: first, actions to create and retain customers; second, actions geared to prevent competitors from dislodging a company from its market position by seizing market share and customers.

Third, *lower-level strategy—or tactics.* This level requires a shorter time frame from those at the two higher levels. Normally, it links with a company's or product line's business plan, marketing plan, and annual budget.

In everyday application, tactics are actions designed to achieve short-term objectives, while in support of longer-term objectives and strategies. Also, tactics are precise actions that cover such areas as social media, the Internet, sales force deployment, supply-chain methods, customer relationship programs, training, product branding, value-added services, and the selection of a market segment to launch a product or dislodge a competitor.

The Strong versus the Weak

Within the framework of applying military strategy to business practice, another viewpoint stubbornly exists: If only the wealthy firms can afford advanced technologies and are able to sustain an ongoing financial commitment, are they more likely to enjoy an overwhelming competitive edge?

Is it also assumed that smaller, cash-strapped companies with fewer resources are always at a disadvantage? Do they have to defend their markets and retain customers against the continuing threat of resource-rich organizations that can mount offensives and win over markets on their own terms?

The long history of conflicts, however, challenges those assumptions. By examining the chronicles of military encounters from ancient times to the present day, there is substantial evidence about groups clashing in strong versus weak battles where the weaker side emerges as victor in the proverbial David versus Goliath encounter.

Thus, the resulting lessons that span 2,500 years of conflict-filled history have merged into firmly established concepts, guidelines, and rules. They provide insight into how leaders in those uneven struggles operated with comprehensive strategy plans and implemented them through effective direction.

The subsequent body of knowledge survived meticulous analysis by scholars and practitioners and has been translated into numerous languages. This knowledge, in turn, has been incorporated into the curricula at universities and military academies worldwide. And these same time-tested ideas, concepts, and rules continue to flourish today in business books, magazines, and workshops.

So it is that military theory is interwoven into business plans, cited at product-line strategy meetings, and discussed at formal strategic sessions in the C-suite. As the masters of strategy say:

> Money is not the sinews of war although it is generally so considered. It is not gold, but good soldiers that insure success.
>
> **Machiavelli**

> If we always knew the enemy's intentions beforehand, we should always, even with inferior forces, be superior to him.
>
> **Frederick the Great**

> Supreme excellence consists in breaking the enemy's resistance without fighting.
>
> **Sun Tzu**

Transposing Military to Business: A Personal Discovery

I first realized the parallels between business strategy and military strategy several decades ago when I came upon two books that have commanded a central position on my bookshelf throughout my career: Sun Tzu, *The Art of War* (Samuel B. Griffith's translation) and *Strategy* by B. H. Liddell Hart, the renowned twentieth-century British historian.

Once smitten, I found other military books and began transposing the concepts into my first business-related article, "Corporate Objective and Marketing Aim: What Is the Relationship?"* The following applications come from that article:

1. The more strength you waste when entering a market, the more you increase the risk of the competition turning against you. Even if you succeed in winning the market, the less strength you will have to profit by the victory.

---

* Paley, Norton. *California Management Review* (University of California), Winter 1968, p. 59.

2. While the tools of marketing—advertising, sales promotion, field selling, market research, distribution, etc.—are physical acts, their direction is a mental process. The better your strategy, the easier you will gain the upper hand and the less it will cost you.
3. The tougher your marketing practices, the more bitter you will make your competition, with the natural result of hardening the resistance you are trying to overcome, which tends to consolidate all the competition against you.
4. Following the above point, the more intent you are in securing a market entirely of your own choosing and terms, the stiffer the obstacles you will raise in your path, and the more cause you will provide for others to try to reverse what you have achieved.
5. Where you are too evenly matched with the competition to offer a reasonable chance of early success to either, it is wise to learn from the psychology of strategy that, if you find your competition in a strong market position that will be too costly to dislodge, you should leave him a way out of the market—as the quickest way of loosening his hold.

From that starting point, I transposed more military concepts and showed their applications to business operations in my numerous books, articles, and seminars over the next 30-plus years.

This book, then, is a meticulous selection from my collective writings and is organized to blend historical lessons with modern business practices. I have overlaid these writings with a new level of thought that is applicable when strategizing to win in today's embattled marketplace.

It will serve you well to internalize and practice the ideas, for their enduring wisdom provides a solid platform to understand, develop, and apply competitive business strategies to your own situation.

In sum, you will find in the following chapters how strategy works, how to develop a competitive strategy, and how it is implemented through a strategy action plan.

After reading and internalizing the lessons and applications in this book, you should be able to do the following:

1. Define the historical roots of strategy and apply the concepts to today's global and unrelenting competition—whether you manage a small business, a mid-sized company, or a multi-divisional enterprise with global market coverage.

2. Apply business strategies to turn your risky competitive situation into a fresh market opportunity.
3. Align the goals of the business plan to your organization's culture and thereby improve the chances for its successful implementation.
4. Personalize a leadership style to maximize your performance.
5. Think like a strategist to outthink, outmaneuver, and outperform your competitors.
6. Utilize a strategy diagnostic system to evaluate your strategies.
7. Develop a strategy action plan.

ORGANIZATION OF THE BOOK

This book has nine chapters, the detailed contents of which are outlined below. In addition, three appendices provide the reader with strategy tools.

Appendix 1 describes a strategy diagnostic tool. This tool helps assess a firm's competitive strategies and provides a reliable performance measure to gauge the likelihood of success, before committing resources.

Appendix 2 is titled "Appraising Internal and External Conditions." This checklist assists in analyzing those key factors—markets, customers, and competitors—that affect your ability to carry out your business plans. It also helps you assess your company's internal competencies.

Appendix 3 provides a strategy action plan that includes forms and guidelines for you to use to develop a business plan based on a proven format. You can personalize the plan by inserting specific terminology and issues related to your firm.

Clausewitz observed that "in war the result is never final." The same can be said of the study of the classical works on strategy presented in this book. Like other classics in such fields as art, philosophy, and religion, they are open to different interpretations according to the background, interests, and perspective of *you*, the reader.

Thus, in this book, "the result is never final." That is, there can be no absolute, definitive interpretation or understanding of the particular passages I selected for interpretation and application to competitive business strategy. Accordingly, I suggest an interpretation, not a dogmatic point of view. Given the content, you can formulate your own points of view about how the strategies would apply to your particular situation.

One final note about the quotes used throughout the chapters: You will gain valuable insights from the military masters of strategy if you simply substitute terms, such as "enemy" with "competitor," "war" with "competitive confrontations," "battles" with "campaigns," "general" with "senior executive," and so on.

Good reading and good luck.

Chapter 1. Apply Strength against Weakness: Maneuver by Indirect Strategy.

The indirect strategy endures as the indomitable principle that consistently stands out as one of the vital ingredients of a business plan. It operates in three dimensions: First, an indirect strategy is anchored to a line of action whereby you apply your strength against a competitor's weakness. Second, as you activate an indirect strategy against a competitor, focus on serving customers' needs or resolving their problems in a manner that outperforms your competitor's actions. Third, your aim is to achieve a psychological advantage by creating an unbalancing effect in the mind of the rival manager. That is, by means of distractions and false moves, you make it appear that you are launching your effort so as to match your competitor's strengths, whereas your true purpose, as indicated in the first point, is to target his vulnerabilities.

Chapter 2. Improve Chances for Securing a Competitive Lead: Act with Speed.

There are few cases of overlong, dragged-out campaigns that have been successful. Exhaustion—the draining of resources—damages more companies than almost any other factor. "Without exception, all of my biggest mistakes occurred because I moved too slowly," declared John Chambers, CEO of Cisco Systems. Extended deliberation, procrastination, cumbersome committees, and indecisiveness are all detriments to success. Drawn-out efforts often divert interest, diminish enthusiasm, and damage morale.

Chapter 3. Secure a Competitive Advantage: Concentrate at a Decisive Point.

The aim, here, is to gain superiority in selected areas of your market. Therefore, concentrate your resources at decisive points. You thereby emerge stronger than your competitor in key segments of your choosing. Expressed another way: Concentration is your best approach rather than spreading your resources too thin, which will only add more areas of weaknesses by exposing additional points of vulnerability to your competitor. Thus, what matters most is not creating an absolute numerical advantage. Rather, it is developing superiority at a decisive point in the marketplace with the right marketing mix.

Chapter 4. Create a Lifeline to Business Strategy: Employ Competitor Intelligence.

If you know your rival's plans, and you are able to monitor his moves, then you can estimate with some degree of accuracy which strategies are likely to succeed and reject those with minimal chances of success. You can also assess the level of risk associated with each of your options. Further, with good competitor intelligence you can define your competitor's operating patterns. You can determine his market position, positioning of personnel, and where you are likely to face the most or least opposition. Therefore, knowing where your competitor's strength is formidable and where it is weak gives you lead time to take counteractions.

Chapter 5. Maintain High Performance: Align Competitive Strategy with Your Company's Culture.

Corporate Culture is the operating system and nerve center of your organization. It guides how your employees think and react when entangled in a variety of hot spots. A supportive corporate culture drives forward-looking business decisions, generates customer loyalty, and ignites employee involvement. Therefore, learn how to identify the characteristics of high-performing business cultures and see their impact on developing competitive strategies. From there, you can adapt the techniques to reenergize and personalize your company's (or group's) culture.

Chapter 6. The Force Multiplier behind Your Business Strategy: Leadership.

Leadership is about responsibility, accountability, and achieving objectives. Leaders inspire their people, organize actions, develop strategies, and respond to market and competitive uncertainty with speed and effectiveness. Above all, leaders act to win: to win customers, to win market share, to win a long-term profitable position in a marketplace, and to win a competitive encounter before a rival can do excessive harm.

Chapter 7. Engage Heart, Mind, and Spirit: Create a Morale Advantage.

This principle is about the human side of competitive strategy. Success in competitive encounters is an issue of morale, discipline, and trust. In all matters that pertain to an organization, it is the human heart that reigns supreme at the moment of conflict. The level of morale is a gauge of how people feel about themselves, their degree of participation in a team effort, and the confidence they show in their leaders. The aim, therefore, is to create a morale advantage.

Chapter 8. Turn Uncertain Market Situations into Fresh Opportunities: Move to the Offensive.

One indisputable strategy concept has prevailed with steadfast certainty throughout history: When on the defensive, plan for the offensive. Stated another way: If you act defensively, and passively, to protect a market position from an aggressive competitor, your advantage is temporary. Instead, prepare plans to move to the offensive. The alternatives—stalled by lack of ideas, immobilized by fear, standing still, or restrained by blurred imagination—can fester into severe problems. Instead, look at the rule from still another vantage point: when boldness meets caution, boldness wins.

Chapter 9. Think like Strategists: Lessons from the Masters of Strategy.

Studying business history in general and probing past campaigns in particular can sharpen your decision-making and strategy skills. Because no event is a stand-alone occurrence, you can see the whole picture of how one event links to another. And where your skills are strengthened by personal experiences, imagination, intuition, and innovation, you improve your chances of success. You thereby become more adept at uncovering the roots of a problem and determining what went right or wrong.

ANNOTATED BIBLIOGRAPHY

Antisthenes (c. 445 BCE–c. 365 BCE) was a Greek philosopher and a pupil of Socrates. He adopted and developed the ethical side of Socrates' teachings, advocating an ascetic life lived in accordance with virtue. Later writers regarded him as the founder of Cynic philosophy. He was known to have served in the Battle of Leuctra.

Belisarius (c. 505–565) was a brilliant general and the leading military figure in the age of the Byzantine emperor Justinian. As one of the last important figures in the Roman military tradition, he led imperial armies in numerous battles in Persia, North Africa, Italy, and Constantinople.

Carl von Clausewitz (1780–1831) was a Prussian soldier and military theorist who stressed the moral (psychological) and political aspects of war. He was involved in numerous military campaigns during the Napoleonic Era, but he is famous primarily as a military thinker interested in the examination of war. Clausewitz wrote a careful, systematic, philosophical examination of war. The result was his widely acclaimed book, *On War*, which was unfinished at his death. He stressed the conflicting interaction of diverse factors, noting how unexpected developments unfolding under the "fog of war" call for rapid decisions by alert commanders. Known for his maxims, his most famous is that "War is the continuation of politics by other means."

Frederick the Great (1712–1796), king of Prussia, is often admired as one of the greatest tactical geniuses of all time. He was known to have personally commanded his forces from the front in numerous battles, during which times he had six horses shot from under him. When Frederick ascended the throne as king, Prussia consisted of scattered territories. His goal was to modernize and unite his vulnerably disconnected lands, mainly in wars against Austria. During his reign, Frederick succeeded in establishing Prussia as a great European power. Napoleon is known to have said to his officers after the death of Frederick the Great, "Gentlemen, if this man were still alive I would not be here." Frederick and Napoleon are perhaps the most favorably quoted military leaders in Clausewitz's *On War*.

Baron Antoine-Henri de Jomini (1779–1869) was a general in the French and later in the Russian service, and one of the most celebrated writers on the Napoleonic art of war. According to historian John Shy, Jomini "deserves the dubious title of founder of modern strategy." His ideas were a staple at military academies. Prior to the American Civil War, the translated writings of Jomini were the only works on military strategy that were taught at the United States Military Academy at West Point. His ideas shaped the basic military thinking of its graduates.

Sir B. H. Liddell Hart (1895–1970) was an English soldier and military historian. He is best known for his concept of the *indirect approach*, which he viewed as a valid strategy in other fields of endeavor, such as business and sports. He is also famously recognized for two fundamental principles: First, direct attacks against an enemy firmly in position almost never work and should never be attempted. Second, to defeat the enemy, one must first upset his equilibrium, which is not accomplished by the main attack, but must be done before the main attack can succeed. Liddell Hart also claimed that, "The profoundest truth of war is that the issue of battle is usually decided in the minds of the opposing commanders, not in the bodies of their men." From *Strategy* by B.H. Liddell Hart, Frederick A. Praeger Publishers, New York, Washington 1965.

Niccolo Machiavelli (1469–1527) was an Italian historian, diplomat, philosopher, humanist, and writer based in Florence during the Renaissance. He is acknowledged by many as a founder of modern political science. Machiavelli is most celebrated for his book, *The Prince*, written in 1513, which contains maxims on acquiring and keeping political power. Since the sixteenth century, generations of politicians have been attracted, and repelled, by his writings about the immorality of powerful men. The term *Machiavellian* has since become a negative term describing someone who aims to deceive and manipulate others for personal advantage.

Mao Tse-tung (1893–1976) was a Chinese Communist revolutionary, guerrilla warfare strategist, Marxist political philosopher, and leader of the Chinese Revolution. He was the architect and founding father of the People's Republic of China. It was during the "Long March" of 6,000 miles that Mao emerged as the top Communist leader. During the Sino-Japanese War, Mao advocated a strategy of avoiding open confrontations with the Japanese army and concentrating on guerrilla warfare. In formulating his concepts of strategy, he was greatly influenced by the writings of Sun Tzu's *The Art of War*.

Napoleon (1769–1821) was a French military and political leader who rose to prominence during the latter stages of the French Revolution. As Napoleon I, he was Emperor of France from 1804 to 1815. He is best remembered for his brilliant successes, often against numerically superior enemies. Napoleon is generally regarded as one of the greatest military commanders of all time. His numerous campaigns are studied at military academies throughout much of the world. Napoleon's invasion of Russia marked the turning point in his fortunes. His *Grande Armee* was badly damaged in the campaign, from which he never fully recovered after his long retreat from Moscow. Final defeat came at the Battle of Waterloo in June 1815.

Pericles (c. 495 BCE–429 BCE) was a prominent and influential Greek statesman, orator, and general of Athens during the city's Golden Age, the time between the Persian and Peloponnesian wars. It was during the latter war that Pericles made his first military excursion. He had such a profound influence on Athenian society that he was acclaimed as "the first citizen of Athens."

Sun Tzu (c. 544 BCE–c. 496 BCE) was an ancient Chinese military general, strategist, and philosopher who is best known as the author of *The Art of War*. It remains as one of the most influential books on military strategy. In particular, the book grew in popularity during the nineteenth and twentieth centuries and saw practical use in Western society. The work continues to influence both Asian and Western strategy, culture, and politics. *The Art of War* is widely studied in most military academies worldwide and has become increasingly popular among political leaders and those in business management. (Sun Tzu is extensively quoted in this book, along with the writings of Clausewitz.)

Thucydides (c. 460 BCE–c. 395 BCE) was a Greek historian and author. His is most credited for his *History of the Peloponnesian War,* which recounts the war between Sparta and Athens. He is also recognized as the father of "scientific history," due to his strict standards of evidence gathering and analysis. His text is still studied at advanced military colleges worldwide.

About the Author

Norton Paley has brought his world-class experience and unique approach to business strategy to some of the global community's most respected organizations.

Having launched his career with publishers McGraw-Hill and John Wiley & Sons, Paley founded Alexander-Norton Inc., bringing successful business techniques to clients around the globe including the international training organization Strategic Management Group, where he served as senior consultant.

Throughout his career Paley has trained business managers and their staff in the areas of planning and strategy development, raising the bar for achievement and forging new approaches to problem solving and competitive edge. His clients include

- American Express
- IBM
- Detroit Edison
- Chrysler (Parts Division)
- McDonnell-Douglas
- Dow Chemical (Worldwide)
- W.R. Grace
- Cargill (Worldwide)
- Chevron Chemical
- Ralston-Purina
- Johnson & Johnson
- USG
- Celanese
- Hoechst
- Mississippi Power
- Numerous mid-sized and small firms

Paley has lectured in The Republic of China and Mexico and he's presented training seminars throughout the Pacific Rim and Europe for Dow Chemical and Cargill. As a seminar leader at the American Management Association, he conducted competitive strategy, marketing management, and strategic planning programs for over 20 years.

Paley's published books include *The Marketing Strategy Desktop Guide*, 2nd Edition; *How to Develop A Strategic Marketing Plan*; *The Managers Guide to Competitive Marketing Strategies*, 3rd Edition; *Marketing for the Nonmarketing Executive: An Integrated Management Resource Guide for the 21st Century*; *Successful Business Planning: Energizing Your Company's Potential*; *Manage To Win*; *Mastering the Rules of Competitive Strategy: A Resource Guide for Managers*; and *Big Ideas for Small Businesses*. Paley's books have been translated into Chinese, Russian, Portuguese and Turkish, and his byline columns have appeared in *The Management Review* and *Sales & Marketing Management* magazines.

On the cusp of the interactive movement, Paley developed three computer-based, interactive training systems: The Marketing Learning Systems; Segmentation, Targeting & Positioning; and The Marketing Planning System.

1

Apply Strength against Weakness: Maneuver by Indirect Strategy

History shows that rather than resign himself to a direct approach a Great Captain will take even the most hazardous indirect approach. He prefers to face any unfavorable condition rather than accept the risk of frustration inherent in a direct approach.

B. H. Liddell Hart

The central idea underlying the indirect approach is that you avoid a direct confrontation with competitors. So that, "rather than accept the risk of frustration inherent in a direct approach," your strategy should circumvent your competitors' strong points of resistance. At the same time, your aim is to serve the current and evolving needs of your customers.

This convergence, known as the *two zones of activities*, forms the operational focus by which you (1) devise competitive strategies that outthink, outmaneuver, and outperform rivals threatening to frustrate your efforts; and (2) establish ongoing customer relationships in markets that represent the best opportunity for profitable growth.

Zone one, competitive strategies, utilizes the strongest elements of your marketing and other organizational forces (your competitive advantage) against the weaknesses of your competitor. The resulting strategies make extensive use of (1) market intelligence to estimate the competitive situation, (2) a highly flexible organization or group to respond quickly to sudden threats from rivals, and (3) a competent leadership able to manage financial, human, and material resources to secure and defend a market position.

Zone two pinpoints selected markets and segments that are emerging, neglected, or poorly served. They should be the ones that represent your best opportunity to concentrate your efforts for maximum impact, without exhausting your resources. Consequently, these markets are the starting point and finishing point that precede and follow any actions; they form the epicenter of all your activities.

> The direct and the indirect lead on to each other in turn. It is like moving in a circle—you never come to an end. Who can exhaust the possibilities of their combinations?
>
> **Sun Tzu**

MANEUVER BY INDIRECT STRATEGY

With so much at stake and with so many internal and external forces at play when entering a market and confronting competition, the overriding approach that addresses the numerous market variables and competitive obstacles is the ability to maneuver by the indirect approach. It is the one prevailing guideline that is most likely to determine if the organization grows and prospers, or if it languishes as an also-ran. Thus, where maneuverability is needed and where it is skillfully applied, "who can exhaust the possibilities of their combinations?"

Maneuvering by indirect strategy operates along five dimensions:

First, the strategy consists of a series of actions whereby you apply your strengths against a competitor's weaknesses. The essence of the move is that you position your resources so that your rival cannot, will not, or simply lacks the capability and spirit to challenge your efforts.

Second, concurrent with activating indirect moves against a competitor, your focus should be directed toward serving customers' needs and resolving their problems in a manner that visibly outperforms those of your competitors.

Third, your aim is to achieve a psychological advantage by creating an unbalancing effect in the rival manager's mind, whereby he or she vacillates in indecision. The intent is to disorient and unbalance the competing manager into wasting time and making costly and irreversible mistakes. Using distractions and false moves make it appear that you are launching your effort directly at the competitor's strengths, whereas your true purpose is to target his or her vulnerabilities.

Fourth, make every effort to convince your competitor that continuing aggressive efforts will be too costly, with little or no chance of justifying the expenditures of people, money, and materials.

Fifth, fighting in the marketplace is not your intention. Rather, your aim is *possession*. That is, your purpose is to hold a long-term position in a target market, as gauged by attaining a market share objective, securing a position on the supply train, reaching a profitability goal, or similar metrics.

> The concept of war does not originate with the attack, because the ultimate object of attack is not fighting; rather, it is possession.
>
> **Clausewitz**

All five applications serve the strategic purpose of reducing any resistance leveled against you. You can then utilize the full power of your resources without squandering them on strength-draining actions. From another viewpoint, the indirect strategy is an encounter of manager against opposing manager: your experience and skills pitted against those of your opponent.

The advantages and applications of the indirect strategy are illustrated by the following benchmark examples from past decades:

German and Japanese automakers first entered North America with small cars during the energy crisis. They shrewdly and skillfully avoided a direct confrontation in a market essentially neglected by domestic manufacturers during the 1970s, and poorly served during the 1980s and 1990s. Once embedded, they prudently expanded into full lines of cars covering all price segments of the market, with resounding success.

Miller Brewing Company correctly identified the light-beer category as an emerging market, which evolved into the largest segment of beer drinks, with Miller Lite™.

Dell Computer started out by bypassing the traditional distribution channels through retailers and other intermediaries. Instead, the company sold directly to the end user with a build-to-order strategy, which complemented its low-price approach.

Apple Computer became a dominant factor in schools early on, specifically occupying that segment, which was left vacant by *IBM*, with computer hardware and software.

Wal-Mart originally opened its stores in towns with populations under 15,000, which were totally ignored at that time by the leading retailers.

With the abundance of classic business examples and credible military doctrine as evidence, it is now safe to conclude that there is never any justification for you, or any manager, to undertake a direct frontal approach by applying your strength against an opponent's strength in today's competitive market. Doing so is reckless in time and resources, violates the primary rule of indirect strategy, and rarely achieves its goals.

Further, a direct strategy means confronting a stronger competitor head-on where there is little or no differentiation in product features, quality, performance, and service; and where there is no perceived advantage in price, promotion, distribution, technology, leadership, or caliber of personnel.

A company that pursues a direct confrontation could end up with severe losses; in other words, a company that attempts to move head on against a competitor actively defending a market position can exhaust itself without reaching its sales, market share, or profitability goals. Should a company achieve some minor objective through a direct effort, such as scoring minimal sales or nominal market share growth, remaining resources will be inadequate for that company to move forward and secure enough market share to reach profitable levels.

> He who knows the art of the direct and the indirect approach will be victorious. Such is the art of maneuvering.
>
> **Sun Tzu**

DEVELOPING AN INDIRECT STRATEGY

There is a logical and systematic process you can use to develop an indirect strategy:

1. Define your strategic goals.
2. Determine the resources needed to achieve your goals.

3. Gather competitive intelligence.
4. Establish security.
5. Implement the indirect strategy.
6. Develop a post strategy.

Define Your Strategic Goals

What does the next 36 to 60 months look like for your market? Based on the best information you can gather, what are the strategic goals or direction of your organization, business unit, or product line? Also based on the most reliable data available, what are specific objectives* for your operation?

As you develop your objectives, make certain to distinguish between the direct and indirect approach. The intent is to outmaneuver competitors and avoid dissipating your strength in misguided directions and costly competitive confrontations.

For instance, during one point in the ongoing battle for a favorable position in the raging mobile phone market, *Motorola* set an objective to regain momentum after relinquishing leadership to Nokia. Most notably, Motorola launched its Razr™ phone, which caught Nokia and other competitors by surprise and captured the attention of a worldwide audience.

The indirect strategy came about when Motorola executives decided to buck the growing industry trend to load phones with cameras and stereo speakers that made them heavier and bulkier. Instead, Motorola introduced a half-inch-thick phone with sleek lines and a shimmering keypad. At the time, Razr did more than just ring up sales; it achieved the strategic goal of getting customers around the world looking again at Motorola as creative, cool, and sexy.

To assist in developing your strategic goals, the following questions will guide you in distinguishing between the direct and the indirect approaches.

---

* These are specific objectives for the period covered by your business plan and include quantitative statements with projections of sales, market share, and other long- and short-term objectives required by your firm. Also included are qualitative objectives related to what you want to achieve in such areas as: new product development, product quality, customer satisfaction, technology, market position, and the like. More details are provided in Appendix 3.

What Are Your Organization's Distinctive Strengths or Areas of Expertise?

Here is where you look at your organization's or business unit's distinctive competencies, such as:

> Competitive strengths of your product or service based on such criteria as customer satisfaction, positive image, and long-term outlook for the brand
> Depth of relationships with intermediaries along the supply-chain
> Efficiency of existing production capabilities
> Availability of finances to fund day-to-day operations and carry out long-term plans
> Commitment to ongoing product development, as well as applications of new technologies
> Quality of customer or technical services
> Level of morale, training, and discipline of personnel, especially of those who interact directly with customers
> Caliber of leadership

The ultimate in disposing of one's troops is to be without ascertainable shape. Then the most penetrating spies cannot pry in nor can the wise lay plans against you. It is according to the shapes that I lay the plans for victory, but the multitude does not comprehend this.

Sun Tzu

What Business Should Your Organization Be in over the Next Three to Five Years?

Here is where you pinpoint the market segments or categories of customers you are likely to serve.

Highlight any standout industry and customer trends that would connect your strategic goals to customers' needs and wants—yet be watchful that "the most penetrating (competitors) cannot pry in nor … lay plans against you." Then you can comfortably elevate your thinking into new product development, instead of relying primarily on the longevity of existing products to sustain company growth.

By taking the long-term view, you begin thinking strategically about how to position your business for the future. In turn, that view determines the breadth of existing and new product lines, which helps you identify new market opportunities. If you are too narrow in defining your strategic goals, the resulting product and market mix will be generally narrow and possibly too confining for growth.

On the other hand, defining your business too broadly can result in spreading capital, people, and other resources beyond the capabilities of your organization. Therefore, look to create a comfortable balance by positioning your business somewhere between the two extremes.*

What Segments or Categories of Customers Will Your Company Serve?

Customers exist at various levels in the supply-chain and in different segments of the market. At the end of the chain are end-use consumers with whom you may or may not come in direct contact.

Other customers along the chain serve as intermediaries and typically perform several functions. They include distributors who take possession of the products and often serve as a warehousing facility. Still other intermediaries repackage products and maintain inventory-control systems to serve the next level of distribution. And there are value-added resellers who provide customer service, technical advice, computer software, or educational programs to differentiate their products from those of competitors.

Examining the existing and future needs at each level of distribution helps you project the types of customers you want to target for the three- to five-year period covered by your strategic goals. Similarly, you will want to review various segments and target those that will provide the best opportunities over the planning period.

What Additional Functions Are You Likely to Fulfill for Customers as You See the Market Evolve?

As competitive intensity increases worldwide, each intermediary customer along the supply-chain is increasingly pressured to maintain a comparative advantage. This question asks you to determine what functions or capabilities are needed to solve customers' problems.

---

* Some individuals advocate reaching far to the extreme and seeking entirely new horizons. Bold, audacious goals are commendable. For those robust managers willing to make such a push, it's prudent to do an in-depth reality check and determine if the corporate culture is oriented to the totally new and untried, personnel are skilled and up to the diverse tasks, and enough resources are available to sustain the effort.

More precisely, you are looking beyond your immediate customers and reaching out further along the supply-chain to identify those functions that would solve your customers' *customers'* problems. Such functions might include providing computerized inventory control, after-sales technical support, quality control programs, just-in-time delivery, or financial assistance.

What New Technologies Will Your Firm Require to Satisfy Future Customer Needs?

Look again at the previous question and think about the practices of your industry. Examine the impact of technologies to satisfy your customers' needs.

From the viewpoint of which areas represent potential indirect strategies, look at where your company ranks with the various technologies and types of software used for product design, manufacturing, and distribution systems. Look, too, at the continuing changes in information technology and business intelligence with the resulting effects on product innovation and market competitiveness.

Also appraise such current technologies as expert diagnostic systems, dashboards, and other business performance management (BPM) systems for problem solving. And look at the rapidly changing communications systems to manage and protect an increasingly wireless enterprise.

What Changes Are Taking Place in Markets, Consumer Behavior, Competition, Environment, and the Economy That Are Likely to Impact Your Company?

This form of external analysis permits you to sensitize yourself to those specific issues that relate to your business from which you can develop an indirect strategy. It is an intimidating task, however, to stay apprised of the vast amounts of information in all of the above areas. Therefore, it is in your best interest to set up (or upgrade) a communications hub that gathers, sorts, and disseminates key news and intelligence through your internal network.

> Generally in battle, use the normal forces to engage; use the extraordinary to win.
>
> **Sun Tzu**

To further illustrate strategic goals and show the application of the "normal" and "extraordinary" forces of indirect strategy, the following case illustrates how a well-known organization engaged a market leader firmly entrenched in a specialized market.

Wal-Mart Stores decided to attack one of the largest consumer-electronics chains, *Best Buy Stores*. Wal-Mart began by sprucing up the interiors of many of its electronics departments and adding several high-end products from companies like Sony, Toshiba, and Apple.

However, simply adding products to its line was "normal" and not a strong-enough move to unseat the leader. To win with an indirect strategy, Wal-Mart would have to do something "extraordinary" by locating and striking at an area of Best Buy's greatest vulnerability.

The area of vulnerability at that time lay in the most profitable line of business for Best Buy: *extended warranties*. Real earnings were not in the sale of the electronic gadgets. Rather, they were in the sale of multiyear protection plans that were actively hawked by the retailers' salespeople. For Best Buy, warranty sales accounted for more than a third of its operating profit.

Wal-Mart pulled out all stops and concentrated on that line of attack by launching extended warranties on TVs and computers at prices that averaged 50 percent below its rival. "Profit on extended warranties has always been the Achilles' heel of Best Buy," declared an industry analyst.

What was Wal-Mart's strategic goal? The company's push into consumer electronics is part of its long-term objective to attract more upscale shoppers. Wal-Mart managers observed that wealthier consumers shopped mostly for food and cleaning products. To get them to go through the whole store, managers reasoned that upgrading the electronics departments as well as other high-end product lines would achieve that strategic aim.

> In all fighting, the direct method may be used for joining battle, but indirect methods will be needed in order to secure victory.
>
> **Sun Tzu**

Strategic aim, therefore, is the first step in developing an indirect strategy that "will be needed ... to secure victory." The purpose, again, is to guide your activities with discipline, rather than wander off in several directions, expending resources without a defined purpose and a measurable end.

There is another but no less important purpose to the strategic aim: to coordinate with the long-term direction of the organization. This broader view adds credibility when seeking budget approval from senior management.

> To accept superiority of numbers as the one and only rule, and to reduce the whole secret … to the formula of numerical superiority at a certain time in a certain place, was an oversimplification that would not have stood up for a moment against the realities of life.
>
> **Clausewitz**

Determine the Resources Needed to Achieve Your Goal

This second step refers to the quantity and type of resources you will need to get the job done. It also determines how you will deploy those resources from a viewpoint of how to maneuver by indirect strategy. Yet keep in mind, as Clausewitz points out, that relying only on the "superiority of numbers" is an oversimplification. Rather, determining the required resources to maneuver by indirect strategy would immeasurably improve the chances of success.

In all respects, shrewd management of all assets is required. As shown in the Wal-Mart example, it is the use of the *normal* and the *extraordinary*. The *normal* relates to all those activities you would normally use to market your products and satisfy customers' needs.

The *extraordinary* points to those unique—and *indirect*—strategies that can significantly outstrip those of your rivals. Usually associated with innovative and differentiated products, value-added services, and other components of the marketing mix, they are difficult for even the most aggressive competitors to imitate—at least for the short term.

Extraordinary examples can cover a variety of categories, as long as they fit the definition and criteria of an indirect strategy. Table 1.1 offers a source of ideas by which you can select extraordinary and normal forces to maneuver by indirect strategy.

> In battle (*competitive confrontations*), there are only the normal and extraordinary forces, but their combinations are limitless; none can comprehend them all. For these two forces are mutually reproductive; their interaction as endless as that of interlocked rings. Who can determine where one ends and the other begins.
>
> **Sun Tzu**

TABLE 1.1

Selecting the Extraordinary and Normal Forces to Maneuver by an Indirect Strategy

Product/Service	Price	Marketing	Supply-Chain	Leadership/Management
Quality	Discounts	Advertising: print, broadcast,	Channels:	Caliber of leadership
Features	Allowances	TV, mobile	E-commerce	Level of employee morale
Options	Payment period	Social media	Direct marketing	Quality of training
Applications	Credit terms	Cross-platform publishing	Distributors/dealers	Managerial competence related to
Style	Special financing	Publicity	Retail	mobilizing resources and
Brand, image,		Personal selling:	Market coverage:	decision-making ability
reputation		Sales force deployment:	Warehouse locations and	Level of expertise in planning and
		Incentives	proximity to customers	developing competitive
		Sales aids	Inventory control	(indirect) strategy
		Samples	and ordering systems	Quality of market and competitor
		Training		intelligence
		Sales Promotion:		
		Webinars		
		Trade shows		
		Events		
		Demonstrations:		
		Sampling		
		Contests		
		Premiums		
		Coupons		
		Manuals		

(Continued)

TABLE 1.1 (*Continued*)

Selecting the Extraordinary and Normal Forces to Maneuver by an Indirect Strategy

Product/Service	Price	Marketing	Supply-Chain	Leadership/Management
Packaging		Telemarketing	Physical transport and	Organizational design related to
Sizes		Internet	timeliness of delivery	internal company
Support services				communications and flow of
Warranties				information through
Returns				organizational layers to field
Versatility				personnel
Uniqueness				Corporate culture related to
Utility				aggressive or passive behavior
Reliability				toward competitors
Durability				
Patent protection				
Guarantees				

Note: To make Table 1.1 into a useful application for your firm, customize the listings in each column by adding or replacing items with those that suit your needs. You will notice that such areas as plant capacity, manufacturing, financial, and technology are omitted. If any fit your definition of direct and indirect, use them. Then select items that qualify as *normal* and *extraordinary* forces to shape an indirect strategy.

The essential point is that you cannot achieve any measure of success without devising strategies that artfully coordinate both "the normal and extraordinary forces." Otherwise, the business suffers the consequences of inching along in a direct, laborious, and resource-draining manner that can only end with marginal or failed performance.

> With many calculations, one can win; with few one cannot. How much less chance of victory has one who makes none at all! By this means I examine the situation and the outcome will be clearly apparent.
>
> **Sun Tzu**

Gather Competitive Intelligence

The third step, and one central to employing an indirect strategy, is the reliability of competitive intelligence to make "many calculations." There is no meaningful way for you to calculate what constitutes *direct* or *indirect*, as well as distinguishes the *normal* from the *extraordinary*, if you do not know the direction in which you are positioned against your competitor. This point is illustrated in the following classic example.

Heublein, the producer of Smirnoff vodka, enjoyed a leading brand position with a dominant market share for two decades. At one point, Smirnoff was attacked on price by a competing brand, Wolfschmidt, then produced by The Seagram Company Ltd.

Wolfschmidt employed a strategy of pricing its product at $1.00 a bottle less than Smirnoff and claimed the same quality. Recognizing a real danger of customers switching to Wolfschmidt, Heublein needed a creative strategy to protect its market dominance. Managers examined a number of options:

1. Lower the price of Smirnoff by $1.00 or less to hold on to market share.
2. Maintain the price of Smirnoff but increase advertising and promotion expenditures.
3. Maintain its price and hope that current advertising and promotion would preserve the existing Smirnoff image and market share.

Although some options were attractive, they were all obvious and mainly direct approaches. Instead, Heublein decided on an indirect strategy.

First, it raised the price of Smirnoff by $1.00 and thereby positioned its flagship product to preserve the premier image, market position, and brand identity it already enjoyed.

Next, Heublein introduced a new brand, Relska, and positioned it head-to-head as a fighting brand against Wolfschmidt's price and market segment. Using that product entry as a means of diverting the opposing manager's attention from other actions, Heublein introduced still another brand, Popov, at $1.00 less than Wolfschmidt.

That action had the decisive effect of enveloping Wolfschmidt by using the normal and the extraordinary. The result was that during one lengthy period, Smirnoff remained number-one in cases of all imported and domestic vodkas shipped in North America, with Popov in the number-two position.

The Heublein case clearly demonstrates how maneuvering operates in the sphere of the indirect approach. It also indicates the indispensable need for competitive intelligence to provide data on the direction the opponent's product is positioned—in this case the relationship of Smirnoff's position to Wolfschmidt's—and to show strategy options for an indirect strategy.

For instance, the physical act of repositioning Smirnoff upscale and then introducing Relska as a threatening fighting brand directly at Wolfschmidt, temporarily distracted the rival manager into inaction.

That move also demonstrates the psychological impact of the indirect strategy on the Wolfschmidt manager. He was sidetracked and unbalanced by the threat to the brand's market share. By indirectly unbalancing Wolfschmidt, the strategy reduced his capability to resist. Further, the total envelopment created by the positioning of the three products caused an inner paralysis that further reduced any action by the Wolfschmidt manager.

> To fight and conquer in all your battles is not supreme excellence. Supreme excellence consists in breaking the enemy's resistance without fighting.
>
> **Sun Tzu**

Not only is competitive intelligence essential to developing strategies, it also helps in identifying the relevant buying behavior of your markets and customers. As important, it reveals valuable information about the pattern of competitor's actions under a variety of market conditions. You accomplish this, in part, through the proliferation of sophisticated data mining programs available for most industries. (More on competitive intelligence in Chapter 4.)

Surfacing in the mid-1990s, data mining evolved to become a vital component in shaping business strategies. If used properly, data can break the competitor's "resistance without fighting." Additional breakthroughs over the past few years have also surfaced for obtaining business intelligence. These are built around innovative behavioral-based programs, as well as vast improvements in analytical tools.

For instance, retailers can now go beyond relying on consumer surveys, which often suffer from inconsistencies about what consumers say they will do and what they actually do. Such retailers as *Montblanc*, *T-Mobile*, and *Family Dollar Stores* are finding new uses for old tools such as in-store security cameras to get behavioral information in real time.

The approach sorts out which variables affect a purchase. Then, armed with the data, managers move rapidly to deploy their salespeople, change color displays, or alter merchandise assortments. Other applications include monitoring shoppers' behavior with devices that track mobile-phone signals. Such data then becomes the foundation for new product offerings that are built around evolving customer behavior.

In the everyday application of competitive intelligence, you can benefit by actively tuning in to market conditions, following buying patterns, and interpreting competitors' moves, thereby providing meaningful options on how to maneuver resources.

To trigger your thinking and assist you in devising an indirect strategy, here are several applications of intelligence:

American Express gathers existing customer information from its call centers and uses the data to make highly targeted cross-sell and up-sell offers to customers.

Charles Schwab compiles routine requests from its investor accounts to form a comprehensive profile of its customers. The in-depth data are then reconfigured and applied to such revenue and profitability goals as customer retention, cross-selling, and up-selling. The aim is to maintain an advanced level of differentiation in an industry that has become intensely competitive.

Hilton Hotels probes for guest information from its hotels and resorts and makes it available in usable form to hotel managers over the Internet. Those managers can then develop new or improve existing services for their guests, and manage corporate loyalty programs and marketing campaigns.

Verizon leverages its data to develop profiles for each of its telecommunications customers. Then, based on each customer's history and preferences, it offers products and services that are likely to appeal to each individual.

In all of the above applications, managers used the hard data to form a two-directional indirect strategy: First, they determined the buying patterns within their respective markets and used the information to devise products, services, and positioning strategies. Second, they incorporated intelligence acquired about their competitor's product and service offerings to shape their own indirect strategies.

> He who wishes to snatch an advantage takes a devious and distant route and makes of it the short way.
>
> **Sun Tzu**

Establish Security

In step four, make certain that your market positions are secure. Security is particularly vital as you concentrate your resources on a specific segment of the market.

The dramatic example of such a breach in security is the case of *Xerox*, which still serves as a constructive lesson for today's challenging competitive conditions. During the 1970s, the company concentrated almost entirely on selling its large copiers to big companies.

Whether by choice or by management's myopic view of the total marketplace, the company was blind-sided as it attempted to protect its dominant market position. In so doing, Xerox managers left a huge gap for enterprising Japanese copier makers to "snatch an advantage (and take) a distant route" by entering an unattended market with their small copiers aimed at the huge number of small and mid-sized companies.

Once established, aggressive rivals such as *Canon* and *Ricoh* took the natural route by expanding their lines of copiers, which finally encroached on Xerox's hold of the large-company segment. The subsequent precipitous drop in Xerox's market share has taken decades to recover. Meanwhile the intruders embedded themselves solidly in the North American market.

Consequently, it is in your best interest to maintain ongoing intelligence about those competitors that might make an indirect attack against you.

The attack could be an unguarded gap in your product line, a flawed service capability, a poorly served user segment, or an overlooked geographic area.

The broader aspects of competitive intelligence procedures are reviewed above. However, do not overlook using the eyes and ears of those involved in one-on-one contact with local customers and competitors. For instance;

Sales reps should observe such local activities as competitors' new product introductions, promotional incentives, or aggressive pricing deals. They should even provide input about their counterparts' morale and level of selling skills.

A *sales manager* can look from a higher vantage point to observe how competitors' salespeople are deployed, types of sales aids used, forms of communication, and methods of compensation. The sales manager would also look at the patterns of customer movements within a defined territory, such as changes in sales and product usage from previous periods.

Advertising and *market research* people can interpret a variety of signs, such as how a new competitor deviated from its usual media and promotional mix, details about impending acquisitions or divestitures, new product developments that might signal an entry into your segment, and any significant shifts in the competitors' key managerial positions.

Financial or *accounting* personnel should examine the financial health of a competitor by looking at such indicators as debt-equity structure, inventory turnover, capital resources, credit rating, and any other clues that would highlight strengths and vulnerabilities.

Production personnel should look for signs that reveal production processes, plant locations and logistics, age of equipment or technologies, quality control systems, and similar areas from which an accurate assessment can be made.

Specifically, here is what you want to find out to strengthen security—and what you would like all those individuals who interface with competitors to uncover:

What is the competitor's strategy for expanding out of its present market segments? Does it have sufficient manpower resources? Would it place your market position in jeopardy?

What is the competitor's financial picture? Can it hold out in a prolonged
   competitive campaign?

What new products, technologies, or services are under development
   that would constitute a threat?

How is the competitor organized and staffed? What is the level of
   morale, discipline, training, and overall preparedness of its staff?
   Are there weaknesses you can exploit?

What is the caliber of the competitor's leadership? Is the leadership
   skilled and experienced? Would their behavior tend to be bold,
   timid, or passive if challenged?

Where are the competitor's potential areas of vulnerability by product
   depth, product quality, service, price, distribution, and reputation?
   (Also review Table 1.1 to conduct an in-depth comparison.)

One mark of a great strategist is that he fights on his own terms or fights
not at all.

Sun Tzu

Implement the Strategy

In step five, your purpose in implementing an indirect strategy is to place
you in a market situation so that, as Sun Tzu advises, "he fights on his own
terms or fights not at all." That means, your central aim is to win customers, to win market share, or to win whatever other objectives you set down
in a business plan in a way that does not exhaust resources through direct
competitive confrontations. The following case describes the implementation of an indirect strategy.

Caterpillar Inc. operated a remanufacturing plant that worked on giant
12-cylinder worn-out truck engines. Workers dismantled them and rebuilt
them for renewed lives of hauling or pushing huge loads. It is hardly
glamorous work to blast away at the buildup of concrete-hard coatings of
grease, dirt, and oil to reach the thousands of parts before rebuilding the
entire engine.

The company got into the business indirectly. It began as a favor
Caterpillar reluctantly did for its customer, Ford Motor Company. To lower
its own costs, Ford's truck-making subsidiary wanted a source to rebuild
engines, which generally sell for half the price of new ones. To maintain
relationships as a supplier to Ford for new engines, Caterpillar consented

and opened a repair shop to accommodate its customer. Where does implementing the indirect strategy come in?

First, the remanufacturing unit listened and responded to a customer-driven request to provide a service. Doing so incurred out-of-pocket expenditures to build a plant to do the low-tech, dirty work of dismantling and rebuilding truck engines. However, it was justified by the long-term strategic aim of maintaining a profitable customer relationship.

Second, Caterpillar's entry into the engine rebuilding business circumvented a recurring problem that existed in its mainline business: dealing in the new equipment part of an industry that is highly cyclical and subject to wide swings in demand due to economic conditions. In contrast, revenues and profits from remanufacturing services continued to climb with some degree of stability. And estimates showed that the segment has excellent long-term growth potential.

Third, Caterpillar management realized that this ancillary business represented a hidden and indirect opportunity to enter another market segment: an under-served and needy segment consisting of a sizeable number of users that could not afford new equipment.

Finally, selling rebuilt products at discount prices formed a solid indirect strategy that blocked makers of knockoff parts out of the lucrative aftermarket and permitted Caterpillar to profit again and again from the same parts.

> The best strategy is always to be very strong: first in general and then at the decisive point. There is no higher and simpler law of strategy than that of keeping one's forces concentrated.
>
> **Clausewitz**

As you implement your indirect strategy, review these guidelines with the intention of "keeping one's forces concentrated" at the decisive point:

Find an unattended, poorly served, or emerging market segment.

Create a competitive advantage by referring to the listings in Table 1.1 and combining them so that they cannot be easily matched by competitors. That means concentrating your strength against the weaknesses of your rival.

Focus all available resources to fulfill the unmet needs and wants of the market in a strength-conserving manner; you thereby solidify relationships with your customers.

The natural goal of all campaign plans is the turning point at which the attack becomes defense. If one were to go beyond that point, it would not only be useless effort, which could not add to success, it would in fact be a damaging one.

Clausewitz

Develop a Post Strategy

In this final step, link a post strategy to your business plan. You are then better able to determine "the turning point at which the attack becomes defense." That point is defined by such metrics as sales, profits, market share, and customer retentions or defections.

There are also warning signals from the marketplace about turning points, such as competitors attacking with claims of technology break-throughs or announcements of superior product quality. Developing a post strategy, then, permits you to set in motion contingency plans to turn an attack into a defense of your market position.

Without a post strategy, your overall business plan is incomplete and can lead to shoot-from-the-hip actions. Further, it places you at the mercy of market and competitive forces with no orderly strategy to extricate yourself from the market, or the ability to turn the situation to your advantage.

A post strategy exists on two levels. First, you can opt to stay and defend the market for the long haul. That choice entails enhancing customer relationships, prolonging the life cycle of an existing product, replacing failing products with new or improved offerings, and managing other areas of the business, for example, service, communications, supply-chain activities, and the like. (Again, refer to Table 1.1.)

Where it comes to prolonging the sales life of a product, some companies have been especially successful in post-strategy planning. One time-tested and noteworthy example is *3M's* Scotch™ brand tape. The product has gone through several cycles spanning several decades and has become an accepted fixture in industrial and consumer markets. The tape has been modified, reformulated, and repackaged for an enormous number of uses, applications, and markets, from heavy factory use, to office, classroom, and home markets.

Second, due to aggressive competitors making it untenable for you to remain in the market, you can suddenly exit the market after ful-filling required obligations to workers, customers, and communities.

TABLE 1.2

Criteria for a Post Strategy to Phase Out a Product

1.	What is the market potential for your product? And what is the product's contribution to your post strategy?
2.	What are the chances of your product being displaced by another product or technology?
3.	What comparative advantage might your company gain by adding value, modifying the product, or creating other differentiating features and benefits?
4.	What would you gain by repositioning your product against the competitor's comparable product to create an indirect strategy?
5.	How much resources (materials, equipment, people, and money) would be available by eliminating the product?
6.	How good are the opportunities to redeploy resources to a new product, market, or business?
7.	What value does the product have in supporting the sale of your other product lines? Is the product line filled out sufficiently to prevent your customers from shopping elsewhere?
8.	Is the product useful in blocking a point of entry against competitors? (See Xerox example.)
9.	What impact will removing the product have on executives' time, the optimum use of the sales force, and relationships with dealers and customers?

Or you may choose to reduce your presence in a market by removing products using a deliberate phasing-out process. (See Table 1.2.)

> Although everyone can see the outward aspects, none understands the way in which I have created victory. Therefore, when I have won a victory I do not repeat my tactics, but respond to circumstances in an infinite variety of ways.
>
> **Sun Tzu**

SUMMARIZING

Three primary characteristics distinguish the indirect strategy:

First, your aim is to always apply strength against your competitor's weakness through actions that "respond to circumstances in an infinite variety of ways." Further, they should be implemented with speed, so that a rival cannot, or chooses not to respond to your actions.

Second, concurrent with triggering an indirect strategy, your intention is to serve customers' needs or resolve their problems with offerings that outperform those of your competitors.

Third, by using speed, concentration, and surprise, your aim is to achieve a psychological advantage by creating an unbalancing effect in the mind of the rival manager. The purpose is to reduce any competitive resistance leveled against you, thereby allowing you to maneuver with the full potential of your resources, so that "none understands the way in which (you) have created victory."

> There are not more than two methods of attack—the direct and the indirect. Yet these two in combination give rise to an endless series of maneuvers.
>
> **Sun Tzu**

Emphasizing an indirect strategy, therefore, is one of the most efficient ways to implement a business plan. Implicit in the indirect approach is that it is possible to confront a larger competitor and win. Thus, avoiding confrontations against a rival and instead relying on "the direct and the indirect" permits you to utilize competitor intelligence to do the unexpected, create surprise through psychological and physical distractions, and move with a comparative advantage.

You then have the advantage of fighting on your own terms, rather than on your competitor's terms. That is, you enjoy a strategic edge by exploiting your own unique advantages through the normal and extraordinary.

2

Improve Chances for Securing a Competitive Lead: Act with Speed

Speed is the essence of war. Take advantage of the enemy's unpreparedness; travel by unexpected routes and strike him where he has taken no precautions.

Sun Tzu

There are few cases of overlong, dragged-out campaigns that have been successful. Exhaustion—the draining of resources—has killed more companies than almost any other factor. "Without exception, all of my biggest mistakes occurred because I moved too slowly," declared John Chambers, CEO of *Cisco Systems*. Thus, an operative approach for you to "take advantage of the [competitor's] unpreparedness" is speed.

Extended deliberation, procrastination, cumbersome committees, and indecisiveness are all detriments to success. Drawn-out efforts often divert interest, diminish enthusiasm, and damage morale.

Additionally, employees become bored and their skills lose sharpness. As damaging, the gaps created through lack of action give competitors extra time to create barriers that can blunt your efforts. Therefore, it is in your best interest to evaluate, maneuver, and concentrate your marketing forces in the shortest span of time.

The proverbial sayings, *opportunities are fleeting* and *the window of opportunity is open,* are especially true in today's markets. Speed, then, is essential for gaining the advantage and exploiting an opportunity. The following example illustrates this point.

An extraordinary opportunity opened for *China Mobile*, the world's biggest phone company by subscribers in 2010. At the time, *Google* faced a painful dilemma: Android™, Google's operating system, ran two-thirds of the smartphones sold in that burgeoning Internet market. Yet Google's online app store, Google Play, was not open for business in China. Google sharply had scaled back its presence on the mainland after quarreling with China's censors.*

China Mobile was quick to exploit Google's absence and rushed in to fill the gap. The carrier immediately launched its Mobile Market store for Android apps and rapidly registered 158 million users. Customers went on to download more than 630 million apps, making Mobile Market the world's largest carrier-operated app store.

As for how speed relates to employees: Beyond the quantitative measures of success, such as sales, profits, and market share, they can see with satisfaction, and perhaps some feeling of pride, that stalled projects move forward, products are launched on time, or market coverage is expanded.

Speed, then, does have an impact on employees; it acts as a unifying element that gives wholeness to managing people and resources. Yet even with the most convincing evidence and far-reaching experiences from such diverse fields as politics, the military, and sports, speed is still largely ignored by many managers.

> Now to win ... and take your objectives, but to fail to exploit these achievements is ominous and may be described as wasteful delay.
>
> **Sun Tzu**

To give pragmatic reality to the strategic value of using speed to "take your objectives," we look at actual marketplace situations through the following six propositions. Doing so should elevate this principle to your conscious level in thinking about how "to exploit [your] achievements."

1. Timing affects market share, product position, and customer relationships—all of which are difficult and costly to recover once given up.

Siebel CRM Systems (acquired by Oracle Corporation in 2005) developed software to manage sales people and call centers. At one point, management saw its revenue nosedive as fierce competition from aggressive rivals blasted in with unrelenting speed and established strong market positions.

"They (Siebel) were basically on a clock. If they thought next year would come along without shareholders seeking alternative board members, they were out of their minds," observed one industry analyst.

---

* As of this writing, all Web content in China is censored, which also affects such websites as Facebook, Twitter, and YouTube.

Siebel was indeed on a clock. Management saw few alternatives other than to move as fast as possible to develop a response strategy before time ran out. Recognizing that competitors would also react rapidly, Siebel did intensify its product development efforts and created a product advantage that it used as the centerpiece for a turnaround.

The strategic point: Recovering lost market share, competitive position, and customer loyalty are often more costly, time consuming, and riskier than moving swiftly at the initial signs of threat. The damage is less severe and the odds better for revival.

> When torrential water tosses boulders, it is because of its momentum. Thus, the momentum of one skilled ... is overwhelming, and his attack precisely regulated.
>
> **Sun Tzu**

2. Where a company stalls and loses momentum, it signals an alert competitor to move in and fill a void.

Motorola Inc. found itself languishing at one point in its mobile phone business. Only through the energetic efforts of a new CEO did the company gain enough thrust to move ahead before its then chief rivals *Nokia* and *Samsung Electronics* could gain market dominance.

Similar to Siebel, the central force behind Motorola's strategy was a rapid push for new product designs. That move entailed more than just instructing designers and engineers to work out the details of a new project. It meant wholesale revamping of the organization's structure, as well as the cultural underpinnings to accelerate the process.

It also meant intrusive, in-your-face leadership by a persistent CEO who incessantly monitored progress, day-by-day, hour-by-hour. Here, too, the clock was ticking for Motorola with continuous and unrelenting certainty.

The strategic point: There are always alert competitors out there probing for weaknesses in a rival company. Or they are searching for a poorly served market segment they can exploit. Try not to give it to them.

3. Speed is a factor in preventing a product from becoming a commodity and possibly causing irreparable damage to a company's reputation.

SanDisk Corporation, the maker of the ubiquitous flash memory devices for cell phones, digital cameras, music players, and handheld game consoles, feared that its product would deteriorate into an indistinguishable commodity.

SanDisk was enjoying a commanding lead with its product line. Revenues surged an average of 70 percent over a three-year period. However, the business was cyclical and market growth began to plateau with existing product designs.

Should competitors get serious about grabbing share from SanDisk by using deep-down pricing, then the market would spin into a profit-draining price war. At one point that painful scenario actually played out when flash memory prices plunged and SanDisk's stock dropped by 30 percent in four days.

SanDisk responded with as much swiftness as it could muster. Certainly management had no intention of sitting and waiting for calamity to hit. To preempt additional attacks from competitors, SanDisk moved with all-out speed to create a brand strategy.

But a brand strategy is more than just hyping the name in advertising. The company needed new products. What followed was a hot-footed effort to pump out new lines, including waterproof memory cards, titanium cards, even memory cards that work only if the rightful owner presses a fingerprint on an embedded reader.

Company managers also moved quickly to solidify customer relationships by developing dedicated products for such key accounts as *Verizon Wireless, Sprint Corporation,* and other wireless carriers.

Strategic point: The inevitability exists that a product moving through the product life cycle will eventually reach maturity as competitors plow in with faster/cheaper/smaller products and commoditize the category. In many cases, you can delay the dire outcomes through continuous improvement and thereby forestall reaching a plateau for longer periods of time.

Product life cycles are a reality and one that certainly has to figure into your business plan. The appropriate time to develop product and market strategies to forestall maturity and sustain advances is during the growth stage of the cycle.

> Any unnecessary expenditure of time, every unnecessary detour, is a waste of strength and thus abhorrent to strategic thought.
>
> **Clausewitz**

4. The risk of losing a viable position in the supply-chain occurs by not moving quickly to secure key customers or middlemen.

EMC Corporation, the maker of storage software and hardware, made a remarkable recovery after the earlier tech bust. It slashed prices, expanded its product line, and repaired relationships with customers.

Through acquisition, licensing, and internal development, EMC moved forward with its strategic objective to help "manage the corporate computing universe." With CEO Joseph M. Tucci at the helm, the company wasted no time in maintaining the momentum by rolling out a series of innovative products and by establishing new partnerships.

To expand rapidly into new markets and secure a foothold before rivals could respond, Tucci lined up *Dell Incorporated* to distribute EMC machines and give it fast access to key markets and customers. Sealing that relationship would secure quicker product adoptions for EMC.

"Customers want to buy from fewer suppliers. We want to be one of the companies they depend on for their enterprise solutions and deny competitors entry," declared Tucci.

Strategic point: Timely and efficient distribution is the bedrock requirement of a successful marketing effort. Losing the position through an "unnecessary detour" within a well-organized supply-chain creates a break that is sure to be filled by a rival firm.

5. Speed adds vitality to a company's operations and behaves as a catalyst for growth. As a major factor toward competitiveness, it impacts virtually every part of the organization.

Cisco Systems and *Procter & Gamble* moved swiftly to secure alliances with important partners early on in China and India. They locked up market positions before their competitors realized what was happening. With enormous growth forecasted in those countries for decades to come, the two companies forged superb positions to reap the benefits from the world's highest growth regions.

In a similar scenario, *Hyundai Motor Company* trounced rivals by driving rapidly into emerging markets. In India, it became a strong number two as it pushed for leadership in the market for small cars. The overall strategy was to move quickly into emerging markets before larger competitors could expand. It is part of Hyundai's strategic goal to become one of the world's largest automakers.

Strategic point: Momentum elevates employee morale and tends to act as a catalyst to energize an entire organization. Certainly, that is a condition any manager would relish.

> Victory is the main object in war. If this is long delayed, weapons are blunted and morale depressed. When troops attack … their strength will be exhausted.
>
> **Sun Tzu**

6. A product strategy that integrates speed with technology and efficiency is in the best position to secure a competitive lead.

Wipro Ltd., a software developer and provider of clients' back-office operations, saw the competitive value of speed and accuracy in handling everything from running accounting operations to processing mortgage applications. In that time-sensitive business, the Bangalore, India, company pioneered low prices and dependability.

However, management assumed correctly that competitors would soon encroach on its growth path. With tasks that were viewed as overly labor intensive, Wipro managers searched for technologies and systems that would automate many of the manual tasks.

Looking at automaker *Toyota*, Wipro managers became entranced with the superb automation and assembly-line techniques that power Toyota's efficiency. Adopting numerous Toyota techniques, Wipro made over numerous business processes with simple, fast, and smooth techniques that replicate the level of assembly-line proficiency achieved by the automaker.

Technology also goes beyond the assembly line: Wipro automated processes to skip manual steps and use analytical software to mine data about their clients' customers. That effort is part of the company's all-embracing move to combine speed, efficiency, and technology improvement within a framework of continuous learning and constant change. It is also part of Wipro's overall strategic objective to become the Toyota of business services.

Strategic point: With victory as the main object, cost control, quality improvement, and customer satisfaction are the ongoing trio for measuring successful performance. It is in your best interest to utilize technology to those ends before your competitor preempts you. Table 2.1 provides guidelines for utilizing speed.

> The whole art of war consists in a well-reasoned and extremely circumspect defensive, followed by a rapid and audacious attack.
>
> **Napoleon**

OBSTACLES TO SPEED

Admittedly, there are stumbling blocks to employing a "rapid and audacious attack." Some are real, others perceived, and any could discourage managers from acting with any measure of bravado. Regardless of source,

TABLE 2.1

Guidelines for Utilizing Speed

Responding rapidly to a competitor's aggressive actions reduces potential losses
 and helps regain momentum.

Drawn-out efforts divert interest, diminish enthusiasm, and depress morale.

Speed is a contributing factor in preventing a product from becoming a commodity.

Speed is a catalyst for growth.

Speed does not mean carelessness, recklessness, or shoot-from-the-hip reactions.
 It requires preplanning, excellent internal communications, and rapid implementation
 of strategies.

Despite the inevitable uncertainties of market behavior, speed is the prudent approach
 to prevent the negative outcomes of procrastination.

Discipline, capability, and skill form the underpinnings of speed.

Indispensable to speed and success is an organizational culture that is totally customer
 driven.

they do become genuine obstacles to speed. For example, consider the
following (mostly legitimate) reasons:

- Lack of reliable market intelligence
- Mediocre leadership that stifles timely and significant progress
- A manager's low self-esteem and indecisiveness as deep-rooted
  personality traits
- Lack of courage to go on the offensive, triggered by a manager's
  innate fear of failure
- No trust by managers in their employees' discipline, capabilities, or
  skills—with a resulting loss of confidence about how they will react
  in a tough competitive situation
- No trust by employees in their managers' ability to make correct
  decisions
- Inadequate support from senior management
- Disagreement and open confrontations among line managers about
  objectives, priorities, and strategies
- A highly conservative and plodding corporate culture that places a
  drag on speed
- Lack of urgency in developing new products to deal with short
  product life cycles (see the SanDisk example above)
- Organizational layers, long chains of command, and cumbersome
  committees that prolong deliberation and foster procrastination

- Aggressive competitors that cause fear among employees, damage morale, and result in lost momentum
- Complacency or arrogance as a prevailing mind-set

The above list reflects serious stumbling blocks to speed. Any of them can solidify as rock-hard deterrents to moving forward and executing a business plan with any reasonable chance of success.

At this point, it should be strongly emphasized that speed in no way means carelessness, recklessness, or shoot-from-the-hip reactions. Good estimates, reliable intelligence, and prudent planning are all required procedures. Once completed, however, speed is your ally to implementation.

> When you see the correct course, act; do not wait for orders.
>
> **Sun Tzu**

We now examine the obstacles in more detail for clues so you can "see the correct course [and] act."

Lack of Reliable Market Intelligence

Even if you conduct formal market research, and even if you have a workable business intelligence system, you should still get actively involved in doing your own grass-roots, hands-on intelligence gathering. You will then be more informed and better prepared to move quickly when presented with an opportunity.

For instance, observe changes in the character of your markets. Define your customers by demographic and behavioral characteristics. Also look for any unmet customer needs that would enable you to respond rapidly in the form of products, services, methods of delivery, credit terms, or technical assistance.

Next, maintain an ongoing dialogue with your customers to find out their most troublesome problems and frustrations. Meet with salespeople. Travel with them. Draw them out on what they see happening in their respective markets. You thereby gain from their insights. Creating such a dialogue makes you and them more focused and able to react.

Also watch for competitors' substitute products that could replace your products or services. Examine customer usage patterns. Likewise, observe deviations in regional and seasonal buying patterns. Check, too, for changes from past purchase practices that could translate into opportunities.

Further, look for innovations in selling, especially with the pervasive use of the Internet and new applications of communications technologies. Tune in to current trends in promotional allowances, selling tactics, trade discounts, rebates, point-of-purchase opportunities, or seasonal/holiday requirements. Again, stay close to salespeople for such information.

Where appropriate, examine your supply-chain and look for opportunities to customize services consistent with the characteristics of the segment. Pay attention to warehousing (if applicable) and what could make fertile possibilities to innovate, such as electronic ordering and computerized inventory control systems that link to data mining capabilities.

Finally, search for innovations and product-line extensions to maintain an ongoing presence in your existing markets, or else to gain a foothold in an emerging segment. As illustrated by *Wipro*, harness new technologies that lead to cost-effective and efficient operations. Doing so could make your company more competitive and result in broadening your customer base.

> They are swift to follow up a success, and slow to recoil from a reverse ... they were born into the world to take no rest themselves and to give none to others. You still delay ... let your procrastination end.
>
> **Thucydides**

Mediocre Leadership Stifles Timely and Significant Progress

The manager with little understanding of the market, and lacking in know-how about the types of actions to take in a competitive situation, cannot be an effective leader—or a competent strategist.

Therefore, effective leadership requires that you work with the best information available, avoid procrastination and be "swift to follow up a success"—even where information about your competitor is faulty or sketchy, and despite uncertainties.*

Within that veil of uncertainty, work hard at developing two indispensable qualities to support your decisions: First, even in moments of apprehension, rely on intuitive guidance to find the proper path. Second, elevate your courage and determination to follow your instincts, however faint.

---

* General Colin Powell indicates that 60 percent of the available information should be sufficient where a decision is required and action needed.

> I habitually think of what I must do three or four months ahead; and I always look for the worst.
>
> **Napoleon**

A Manager's Low Self-Esteem and Indecisiveness as Deep-Rooted Personality Traits

Although Napoleon's actions were planned long in advance, he could also adapt successfully to the pressures of the moment and act swiftly and decisively. Although today's managers realize they must be strong minded, for personal and career reasons many are extremely sensitive to the dangers of a flawed decision.

They feel insecure about what is before them. And even should some be in the habit of acting with speed, the more they linger with the dangers of indecision, the more doubtful they become.

Making inclusive estimates beforehand, supported by all the available market intelligence, and thinking about your alternative courses of action, are the best remedies for indecisiveness. You thereby prepare yourself to take clear-thinking action at any given moment.

> It is still more important to remember that almost the only advantage of the attack rests on its initial surprise. Speed and impetus are its strongest elements and are usually indispensable ... to the goal.
>
> **Clausewitz**

Lack of Courage to Go on the Offensive, Triggered by the Manager's Innate Fear of Failure

Courage is the act of determination in a specific situation. It becomes a character trait only if it becomes a mental habit. Intellect in itself is not courage. There are ample numbers of brilliant managers who simply do not have what it takes to recognize that "speed and impetus" are essential elements for timely and appropriate actions.

It is for you to arouse the inner feeling of courage. Then act. You will have to face the critical moments when reason is pushed aside and replaced by the awful feelings that creep into your mind and take control of your actions.

That is where training, discipline, and experience kick in to overcome those feelings. Keep in mind, too, that you are in a contest of mind against mind; your mind pitted against the mind of a competing manager who may be challenged by similar emotions. You want to be the one who prevails and moves forward.

Consider the extraordinary success of *Google Inc.* Much has been written about the bold and iconoclastic approaches that founders Larry Page and Sergey Brin exhibited when they started operations in 2000. Even then, they showed their bravado by going against the sage advice of seasoned consultants and analysts who advised selling out for a mere pittance during those early days.

Industry watchers saw the behavioral patterns surface again as the search giant quietly acquired a mobile-phone software company and began making forays into the instant messaging, Wi-Fi Internet, and telephony businesses. Where, then, does courage come in?

First, to maintain leadership in its existing business, Google has to keep innovating as the likes of Microsoft and Yahoo! actively search for competing market positions. Next, it takes serious doses of courage to enter new markets that are well protected by such giants as *eBay, Motorola, SBC Communications,* and *Verizon.* Initially, Google had to play catch-up in those fields just to get a foothold. Then, it had to find ways to carve out new or unserved market niches, or enhance its position with unique applications of technology.

> The decision can never be reached too soon to suit the winner or delayed long enough to suit the loser.
>
> Clausewitz

Managers' Lack of Confidence in Their Employees' Discipline, Capabilities, and Skills

This factor is most serious. The best laid plans, the most ambitious goals, and the most vibrant business strategies are not going to work with inexperienced employees who lack the essential business competencies. Also, if they do not display an implacable discipline, or if they cannot demonstrate an aptitude for the job, the organization is in deep trouble. As has been amply demonstrated in this and the previous chapter, discipline, training, morale, and skill form the underpinnings of speed.

That ties to yet another contributing factor to support speed and the push for performance: the amount of time an individual stays in one position. As one executive at *Citigroup* explains, "There is kind of a natural evolutionary process where at some point people have been here a long time and they go off and do something else. That's healthy for an organization."

What is behind that statement? First, speed of reaction is needed at the lower echelons—from field personnel through mid-level managers—so that they can adapt to the unexpected with timely actions.

Second, more often than not, success in the marketplace is rarely a one-shot event. Rather, it is a serial process composed of many localized actions. Exploiting market situations depends on the intelligence and initiative of junior managers. For them, "the decision can never be reached too soon." From their point of view, they are the ones at the grass-roots level most prepared to grasp the need for change, even where senior executives are reluctant to move away from their comfort zones.

Third, long training and extended time at one job level may make managers experts in execution, but such expertise is bound to be gained at the expense of fertile ideas, originality, and flexibility—the essential elements for swiftly meeting the day-to-day demands of the marketplace. Junior managers, therefore, should demonstrate those qualities needed for speedy reaction. This is particularly relevant as the lean and mean organizational format takes hold and field personnel begin taking responsibility for on-the-spot decisions.

Consequently, if there is a lack of trust in some personnel, do not lay blame on them entirely. As pointed out above, a certain amount of attrition is desirable to maintain the agility, motivation, and energy to solve problems and identify moneymaking opportunities. Whatever pathway your organization takes, there still is a primary responsibility to maintain an environment where personnel development through ongoing training and discipline are bedrock components for successful performance.

No Trust by Employees in Their Managers' Abilities to Make Correct Decisions

This is a particularly troublesome problem when employees' morale deteriorates and they lose confidence in their managers' ability to deal with what they perceive as a hopeless situation that could threaten their jobs—and even the company.

A semblance of that condition existed at one point at *Hewlett-Packard*. A new CEO walked into the mind-jarring job with the day-one objective of restoring an $80 billion company to its former glory. With new products as the lifeblood of the business, what would it take to fire up H-P personnel to go back to its cultural roots—to innovate and gain customer confidence? This was the same question raised by many of the intelligent, experienced, and generally savvy H-P personnel. They wondered about accepting and trusting the new CEO.

The essential point is that a convincing display of skill, courage, and determination are necessary traits that managers require, if employees' confidence is to be restored and sustained. (See Chapter 6, "Master Leadership Skills.")

Inadequate Support from Senior Management

If communication gaps exist among managerial levels, the result is inadequate command and control. In turn, such a vacuum prevents senior management from providing timely support in such areas as approving additional investment to increase market share, shifting resources to secure a competitive position, or improving a supply-chain network.

In addition, timely support from management is required in the essential areas of technology sharing, new product development, and personnel training. Underlying all these issues, senior management needs to assure that there is an alignment of the company's overall strategic direction with the mid-level business and marketing plans.

> While we have heard of blundering swiftness in war, we have not yet seen a clever operation that was prolonged.
>
> **Sun Tzu**

Disagreement and Open Confrontations among Line Managers about Objectives, Priorities, and Strategies

Where managers of equal rank cannot resolve difficulties independently, it is up to senior management to intervene. In a multiproduct, multimarket organization, disagreement is typical and understandable, as most contentious issues deal with power and the availability of resources.

That means resources must be allocated among many business initiatives, all of which are vying for attention. Inevitably, some plans get short-changed or receive an outright turndown from management. And even the winners suffer the effects of lost time that slow their efforts.

Painfully, the lapse in time may give the edge to an alert competitor, so that their "clever operation [is] prolonged."

At the giant *Microsoft*, for example, there is intense competition among managers to tap its substantial resources. Beyond money considerations, the company's huge divisions often act as rivals, pursuing overlapping technologies and quarrelling over whose codes will prevail in the markets where their products interact.

A Highly Conservative and Plodding Corporate Culture Places a Drag on Speed

This is one of the toughest barriers to speed. Yet it is one that managers must face. If the culture is out of sync with the competitive environment, and if managers are not in a position to change the culture, they must adapt plans to the existing culture with the aim of moving as rapidly as possible.

At *Royal Dutch Shell* a sluggish culture existed at one time due to the joint British and Dutch management structure, which plodded along with two chairmen and two executive committees. It finally took the courage of a new CEO to streamline the organization. One immediate priority was to speed up the overly analytical culture, which made it difficult for the company to land big deals in a timely fashion.

Lack of Urgency in Developing New Products to Deal with Short Product Life cycles

It is no secret that products exist in a short time frame and in a commercial world that clamors for faster/cheaper/smaller. Where that basic knowledge is not fully internalized by management or acted upon quickly, results can be calamitous.

The life cycle issue has gained so much importance that software developers such as *USG, SAP,* and *Dassault Systems* sell product life cycle management (PLM) software. It is a fast-growing segment of the corporate software market and growing faster than other segments.

PLM software allows people to coordinate their work within a specified stage of the product life cycle, whether they are in London, New York, or Shanghai. The system permits individuals to post ideas for new products so that engineers get to work designing three-dimensional prototypes and manufacturers can lay out a new assembly line, complete with every piece of equipment necessary. With virtually no limitations, thousands of people can participate on a single project from wherever there is a Web connection.

> When the enemy gives you an opening be swift as a hare and he will be unable to withstand you.
>
> **Sun Tzu**

Organizational Layers, Long Chains of Command, and Cumbersome Committees Prolong Deliberation and Foster Procrastination

The essential ingredient for an efficient enterprise is simplifying the system of control and, in particular, shortening the organizational layers from the field to top-level executives.

In a small organization the chief executive officer or president is at the helm. He or she is in a unique position to control both policy making and execution. Because decisions do not have to be channeled through others, they are unlikely to be misinterpreted, delayed, or contested. Plans can be implemented with consistency and speed.

In the larger multiproduct firms with more people, products, and additional levels of authority, results may fall victim to a cumbersome, inflexible operation. Individuals in the field often feel that organizational obstacles in the decision-making process prevent moving forward with new initiatives. Consequently, missed opportunities are common, and *go* decisions get stuck for reasons other than the competition.

Even first-line managers think that there are too many people at the staff level or not enough on the job with revenue divisions. The large office staffs and the shortage of line personnel are sources of constant complaint. Much of that condition, however, has been handled over the past decade by downsizing and reducing staff to an efficient lean-and-mean level.

Your own experience may well support the obvious inference that an organization with many levels in its decision-making process cannot "be swift as a hare." This situation exists because each link in the managerial chain carries four drawbacks:

1. Loss of time in getting information back
2. Loss of time in sending instructions forward

3. Lack of full knowledge of the situation by senior management
4. Reduction of the top executive's personal involvement in key issues that affect the availability of resources

Therefore, for greater efficiency and speed, reduce the chain of command. The fewer the intermediate levels, the more dynamic the operation. The result is improved speed and increased flexibility.

A more flexible organization can achieve greater market penetration because it has the capacity to adjust to varying circumstances more rapidly. It can thereby concentrate at the decisive point before its competitors have a chance to respond.

Forming cross-functional strategy teams made up of junior and middle managers representing different functional areas of the organization can further enhance organizational flexibility.

The psychological is to the physical as three is to one.

Napoleon

Aggressive Competitors Can Cause Fear among Employees, Damage Morale, and Result in Lost Momentum

A physical event can create a powerful psychological imprint on an individual's mind. Such impressions, when spread among employees, can magnify their reactions and generate emotional responses far greater than the actual physical event—expressed by Napoleon as a 3-to-1 ratio.

Some feelings are correct, others distorted. That means employees will react to signs they choose to interpret and believe through observation, feel, or through the inevitable rumors—right or wrong.

What follows is that employee morale could flounder based on the slightest signs of what they perceive as a reversal. Such warnings may include failed performance of a new product, reduced profits, changing customer behavior, competitors grabbing key customers, or other forms of perceived disruptions. Thus, what the mind can conjure up, believe, and then react to, becomes an individual's reality.

In addition, there are the sudden changes in management structure, the resignation of a key executive, or the downsizing of operations that can send shivers through the employee ranks. The discipline, training, leadership, communications, and underlying culture of the organization that will greatly influence how employees react to these signs. We now look more closely at each influence.

Discipline

Toughening up employees to withstand the natural swings that exist in a global and competitive market takes a great deal of leadership skill and sensitivity to their attitudes and behaviors. This is particularly so when the natural inclination for individuals is to shrink from the realities of an uncertain environment.

Therefore, your aim is to inspire your staff to make courageous save-my-company, save-my-job efforts and turn potential defeat into victory. Accordingly, maintain a continuing awareness of what the psychological effects of your business strategies can have on your subordinates.

For instance, if individuals are placed in a no-escape predicament and faced with the bleak outcome of dismissal, they are not likely to shrink from self-interest. Instead, they will often use their utmost creativity and energy to fight out of the tight spot.

Thus, a real or perceived threat often creates remarkable behavior among individuals—*if* inspired by an energetic manager with a disciplined strategy for survival, followed by a plan for a burst of growth.

Training

There is the pragmatic truth that if employees are unaccustomed to the rigors of travel and long hours of work, they will worry and hesitate at the moment when level-headed decisions are needed to handle tough competitive conditions. It is here that training and discipline support the investment of time and money.

Leadership

There are particular qualities that define a manager's ability to lead. These include insightfulness, straightforwardness, compassion, courage, and strictness, which are described in Chapter 6.

Communications

Communications refers to the internal structure of the organization whereby timely information is disseminated and decisions are implemented.

This influence was referred to above under barriers created by organizational layers, long chains of command, and cumbersome committees.

The essential point is that clear, uncluttered communications should convey meaningful information, such as objectives about the long-term direction of the organization or group, goals related to strengthening customer relationships, trends in the marketplace, and intelligence about competitors. At the tactical level, information covers such areas as e-commerce, promotion, and customer service objectives.

Culture

Indispensable to acting with speed and decisiveness is an organizational culture that helps all personnel acquire a way of thinking and an orientation that is totally customer driven.

A positive corporate culture drives ambitious business decisions, generates customer loyalty, and ignites employee involvement. It extends to maintaining positive relationships in dealings with customers and suppliers. It is especially seen in the attitudes and reactions by employees when threatened by overly aggressive rivals bent on blocking company efforts, or when attempting to secure a foothold in its market.

> Quick decision is sought in campaigns and battles, and this is true at all times ... a long drawn-out war is considered harmful.
>
> **Mao Tse-tung**

Complacency or Arrogance as a Prevailing Cultural Mind-set

An outlook with a cocky, "we've-got-the-market-locked-up" attitude often translates to complacency and ends with defeat. Such feelings tend to dry up creativity and stifle the inner drive for excellence. It siphons off the adrenalin that comes from fighting within a performance-based arena.

The marketplace is littered with one-time leaders in such fields as automobiles, electronics, metals, and consumer goods. Some companies in these industries turned around at the last moment; others were acquired or went bankrupt. The prevailing mind-sets of key individuals in those losing companies were usually supported by the inaccurate notion that high market share means security—or they were too big to fail.

What generally followed was a passive defense of their respective market positions, whereas a "quick decision is sought in campaigns" to curb further reversals. Those conditions at one time or another were faced by such high-profile companies as *Xerox, General Motors, EMC,* and *Hewlett-Packard.* All, however, have acted aggressively to pull out of the slump or are in the process of doing so.

> When the enemy presents an opportunity, speedily take advantage of it.
>
> **Sun Tzu**

SPEED IN A TWENTY-FIRST-CENTURY GLOBAL SETTING

The global marketplace can appear overwhelming with its diverse cultures, regulations, and government interventions, such as the ones that affected Google in China. In addition, there are the numerous global and local competitors creating challenging situations that often build into serious barriers to speed. Yet, companies do overcome obstacles and create opportunities by integrating their long-term objectives with competitive strategies in the local marketplace.

Such is the case of Paju, a town in South Korea. Factories were being built at breakneck speed by the Koreans, Japanese, and Taiwanese in the fight to own the global television market. In particular, *LG.Philips* allocated $25 billion in Paju to expand its plant eightfold over a few years.

Meanwhile, *Samsung Electronics* spent $20 billion to expand its LCD panel operation. As fast, Japan's *Sharp Electronics* erected a $1.4 billion plant to make large LCD glass panels in other parts of South Korea. This buildup in Korea and beyond has huge implications for the televisions that consumers will be watching over the next decade.

Along with productivity and efficiency come lower prices, followed by the penetration of huge markets globally, which virtually assure those companies' dominance in the television industry over several decades. Beyond television, LCDs are used for millions of computers, and an array of millions of cell phones, personal digital assistants, digital audio players, and cameras.

> If we wait for the arrival of reinforcements, the (opponents) will certainly destroy us. Therefore, before the enemy is assembled we should immediately attack in order to blunt his keen edge and to stabilize the morale of our own troops.
>
> **Sun Tzu**

SUMMARIZING

From virtually every field of endeavor, speed surfaces as one of the most important elements for successfully implementing a strategy. As pointed out, this imperative does not mean reckless or impulsive movement. It does forewarn, however, that once careful and prudent estimates have been made, hesitation and indecision become your enemies.

From a total company viewpoint, and within a global setting, speed impacts a number of managerial, organizational, and competitive issues. Your aim is to move forward immediately to "blunt [your competitor's] keen edge." For instance, it affects such areas as

How you organize your company's or business unit's pecking order so that your decisions flow smoothly without getting stuck in a maze of managerial layers, which can result in distortions and misinterpretations of what was originally intended

How fast you react to a global competitor oceans away, where his clear-cut aim is to feed off your customers and erode your market position in your local marketplace

How confident you are in your business plan, so that strategies and tactics planned in advance are set in motion in a timely manner—rather than react with hasty, fits-and-starts movements

How and when you launch a new product to gain a competitive edge and secure a favorable share-of-mind position among early adopters

How rapidly you harness the exceptional advantages of the Internet and integrate it into your internal operations, so that the technology impacts favorably on customer solutions

How fast you adopt new systems to foster virtual communications within your organization and along the entire supply-chain

There is no instance of a country having benefited from prolonged warfare.

Sun Tzu

In sum, Sun Tzu's sage advice is readily transferable to a competitive confrontation you may face. That is, tardiness would unquestionably affect many of the internal functions of your organization and influence your ability to "stabilize the morale" of your employees. Further, sluggishness, often identified as an ingrained cultural trait of the organization, could (if it exists in your firm) reflect on your managerial style and, in turn, could inhibit your ability to lead under real-time market and competitive conditions. Therefore, you can improve your chances for securing a competitive lead by acting with speed.

3

Secure a Competitive Advantage: Concentrate at a Decisive Point

There is no higher and simpler law of strategy than that of keeping one's forces concentrated ... to be very strong; first in general, and then at the decisive point.

Clausewitz

Finding the decisive point focuses on "keeping one's forces concentrated." It is an essential task that should occupy your thinking as you shape a strategy for your company, business unit, or individual product. Choose correctly and you are likely to win the competitive battle. Choose incorrectly and you could end up spreading your resources thin and slugging it out in the marketplace, which ends up consuming excessive amounts of human, material, and financial resources.

What is a decisive point and how can you find it? It turns out there are numerous possibilities for selecting a decisive point for a concentrated effort. The general guideline is that you target a competitor's specific weakness or a general area of vulnerability.

The search begins by conducting a comparative analysis. This includes reviewing a wide range of factors, from markets and products to corporate culture, that affect the timeliness and efficiency of how a company deals with a competitive situation. Encompassing those factors are the psychological or human aspects of people who must perform the work.

Now if the estimates made before hostilities indicate victory, it is because calculations show one's strength to be superior to that of his enemy; if they indicate defeat, it is because calculations show that one is inferior. With many calculations, one can win; with few one cannot. How much less

chance of victory has one who makes none at all! By this means I examine
the situation and the outcome will be clearly apparent.

Sun Tzu

Table 3.1 lists specific areas for a comparative analysis. It is your starting point to search out a rival's vulnerabilities, as you compare them with
those of your company.* When completed, you are better able to develop
a strategy that singles out a decisive point at which to concentrate your
resources. You are thereby in an enhanced position to "examine the situation and the outcome will be clearly apparent."

As you will see in the table, there are a considerable number of variables
involved for a proper assessment, with the likelihood that you will add
other possibilities unique to your business.

Yet, even with the utmost care, there are bound to be gaps in your
assessment, especially in less visible areas concerning a competitor that
cannot be clearly measured, isolated, or controlled. In addition, there are
the undercurrents of friction, chance, and luck that emerge at unexpected
times to alter your assessment.

The effects of inadequate competitor intelligence further complicate the
process and add to the shifting nature of risk. The human element, with
its unpredictable nature of creativity, talent, intuition, and experience of
those individuals conducting the analysis, also add to the complexity of
conducting an assessment.

> Engage ... only when odds are overwhelmingly in your favor. Always
> identify and exploit your comparative advantage.

Sun Tzu

All these gaps can undermine the objectivity you need to assess your own
company, as well as your competitors. Nonetheless, the search to "identify
and exploit your comparative advantage" and pinpoint a decisive point of
concentration must overcome these hurdles by using a variety of approaches
for comparative analysis—along with the input of your best judgment.

There is just too much at stake to avoid the implications of this irrefutable
lesson of history for the strategy development process, among them to avoid
a head-on confrontation with a competitor, prevent depleting resources,
and forestall a prolonged, dragged-out, morale-draining campaign.

---

* Comparative analysis is also known as net assessment. Note, too, that the content of Table 3.1 has
similar criteria to Table 1.1.

TABLE 3.1

Comparative Analysis to Identify a Decisive Point[a]

Categories for Strength/Weakness Comparisons	Competitor's Strengths and Weaknesses (5 = strong, 1 = weak)	Your Company's Strengths and Weaknesses (5 = strong, 1 = weak)[b]	Identify the Decisive Point and Describe the Comparative Advantage
Markets			
Market segments served and levels of penetration			
Market position, e.g., reputation, market leader, follower, or niche specialist			
Commitment to long-term development of a market; level of investments in people, research, and technology			
Ability to sustain long-term customer relationships			
Product or Service			
Quality			
Features			
Reliability			
Packaging			
After-sales services			
Warranties			
Returns policy			
Level of technology			
Applications			
Brand name awareness			

(Continued)

TABLE 3.1 (*Continued*)

Comparative Analysis to Identify a Decisive Point[a]

Categories for Strength/Weakness Comparisons	Competitor's Strengths and Weaknesses (5 = strong, 1 = weak)	Your Company's Strengths and Weaknesses (5 = strong, 1 = weak)[b]	Identify the Decisive Point and Describe the Comparative Advantage
Stage in product life cycle (introduction, growth, mature, decline)			
Frequency of new product introductions			
Price			
List price			
Discounts			
Allowances			
Credit and financing terms			
Communications			
Advertising commitments (business to business, business to customer)			
Sales force (selling skills, training, sales aids, incentives, compensation, motivation, market coverage)			
Sales promotion (trade shows, webinars, contests, premiums, coupons, other)			
Telemarketing and mobile communications			
Internet (describe by usage and application)			
Publicity			
Market research and competitor intelligence			

Supply-chain

Sales force directed to end-use customers and/or intermediaries along the supply-chain

Market coverage (identify gaps in reaching emerging, neglected, or poorly served market segments)

Inventory control systems

Physical transportation

Support systems along supply-chain

Personnel[c]

Customer-driven orientation vs. product-driven mind-set

Competitive spirit and motivation

Level of morale and group unity

Market awareness and ability to foresee trends

Competitor awareness and ability to deal with threats

Overall skill level by job function

Overall experience and knowledge of the business or industry

Level of understanding about strategy and tactics and ability to apply techniques to market opportunities and competitive threats

The Organization

Profile of company culture, e.g., passive or aggressive (as it relates to market and competitive situations)

Supportive management

Work environment

(Continued)

TABLE 3.1 (Continued)

Comparative Analysis to Identify a Decisive Point[a]

Categories for Strength/Weakness Comparisons	Competitor's Strengths and Weaknesses (5 = strong, 1 = weak)	Your Company's Strengths and Weaknesses (5 = strong, 1 = weak)[b]	Identify the Decisive Point and Describe the Comparative Advantage
Internal communications			
Managers' ability to react to market opportunities and threats			
Managers' competence in planning and developing competitive strategies			
Commitment to ongoing training and development of personnel			
Financial resources (ability to sustain operations)			

[a] The categories are applicable to most businesses and cover many of the key areas for conducting a meaningful comparative analysis. For your own specific needs, feel free to interpret, add, delete, and generally modify the listing to fit your business and industry.

[b] Even flawless intelligence on your competitor is of little use if estimates of your own strengths, weaknesses, and performance are inflated or badly misjudged. Ironically, obtaining accurate and unbiased information about your own resources is often the most challenging aspect for preparing a comparative analysis. This is especially so when assessing such cloudy areas as company reputation, quality of customer service, communications, attitudes, and morale. Nevertheless, make your best estimates by using a team approach.

[c] For your own personnel, you can conduct your comparative analysis by group levels, e.g., senior and mid-level managers, customer support, and field personnel. Even if all the documentation for a finite assessment may not be available to you—and it certainly will not be for a competitor's personnel—you have to rely on reputation, third-party estimates, and your own observation.

Another type of comparative analysis you can use to augment Table 3.1 is the familiar SWOT analysis, shown in Chapter 8, Table 8.1. This analysis is a widely used and time-tested approach, especially within the framework of looking for a decisive point. When employed in a group setting, it provides a highly reliable technique for estimating your internal and external situation. The underpinning for the entire comparative analysis relies on a sound competitor intelligence system, which is detailed in Chapter 4.

> The employment of forces should be regulated by two fundamental principles: first, to obtain by rapid movements the advantage of bringing the mass of the troops against fractions of the enemy; second, to strike in the most decisive direction where the consequences of his defeat may be most disastrous to the enemy, while at the same time his success would yield him no great advantages. The whole science of great … combinations is comprised in these two fundamental truths.
>
> **Jomini**

From a total strategy viewpoint, therefore, concentration provides the means by which you can bring your strength to bear against the vulnerability of your rival. You thereby "obtain by rapid movements the advantage of bringing the mass of [resources] against fractions of the [competitor]."

That point of concentration takes place where you and your competitors interact: within market segments and their subdivisions, niches. Consequently, the remainder of this chapter deals with the epicenter for competitive confrontations, which is the marketplace and more specifically, the decisive points of concentration: market segments.

> It is a major act of strategic judgment to distinguish these centers of gravity in the enemy's forces and to identify their spheres of effectiveness.
>
> **Clausewitz**

DEFINING A DECISIVE POINT OR SEGMENT

What does a decisive point or segment look like? How would you "identify their spheres of effectiveness"? Decisive points can be viewed from a variety of viewpoints. First, from a broad market perspective,

China is a prime example of several spheres or decisive points. For instance, take its enormous population of 1.3 billion, speaking more than 100 dialects. That alone makes China about as diversified as any single country can be from a marketing perspective.

It is with such immense diversity that segmentation plays a central role in developing a viable portfolio of opportunities. In turn, gateways open to what people eat, wear, and drive—from north to south, east to west, rich to poor, young to old, city to countryside. From a manager's position, then, China stands out as a superb example of the potential opportunities when using decisive points to penetrate a market.

A second viewpoint is where a company enters a new market and introduces its product in a target market where its name and product are unknown and the firm has no prior experience. That was the situation Chinese auto maker *BYD Co.* faced when it finalized plans for an all-electric battery car that would take about seven to nine hours to fully charge when plugged into a regular home outlet. The business plan called for building its brand name in North America by offering one model, a five-seat car, before eventually expanding its offerings.

The launch strategy pinpointed a decisive segment within the United States, with projected initial sales of a few hundred units through a small number of dealers and priced at slightly more than $40,000. Defining the segment more precisely, one BYD executive stated, "In the beginning, our target customers were government agencies, utilities, and some celebrities."

A third view of a decisive point is a previously overlooked but substantial group. That was the case with the boomer generation—people born between 1946 and 1964. The boomers were characterized with behavioral patterns and brand choices that were hardened. Further, it was thought that their biggest earning and spending years were behind them.

Those assumptions proved wrong on all counts. Once awakened to the reality that a large and vacant segment was up for grabs, the more savvy managers in a variety of organizations began reexamining the boomers and came up with startling discoveries. Strong evidence revealed that this ignored group was far from sluggish, inactive, and set in its ways. Instead, with average life expectancy at an all-time high in many of the subgroups, individuals over age 50 considered the middle age years as a new start in life and were acting as if they would be around for two, three, or more decades.

Surveys further indicated that few boomers planned to stop working altogether as they aged. With children grown and out of the house, the lifestyles shaped up like this: Second careers started, boomers strived to stay mentally and physically active, and they were open to new experiences and products. Those signals awakened marketers to move rapidly and gain a solid foothold in a previously disregarded segment that now exhibits outstanding promise for long-term growth.

In all these examples, succeeding by focusing on decisive points or segments incorporates a four-step process:

1. Employ market intelligence to define customer segments and evaluate competitors.
2. Conduct a comparative analysis to identify one or more decisive points.
3. Pinpoint a segment for initial entry; once secured, systematically roll out into additional market segments—still using comparative analysis and market intelligence.
4. Customize products and services to those segments with unique and definable benefits.

Employing segmentation or focusing on a decisive point, therefore, means splitting the overall market into smaller submarkets or niches where you enjoy measurable and substantial advantages over a competitor. Then you can concentrate your strengths at decisive points against the weaknesses of your rival.

In turn, that approach translates to strategies you can apply to product differentiation, value-added services, and business solutions that exceed those offered by your competitor. Doing so gives you a significant chance to win over a larger competitor that may be spreading its resources over numerous segments, yet has allocated only modest expenditures to the segment you selected by comparative analysis.

Weigh the situation, and then move. Now the elements ... are first, measurement of space; second, estimation of quantities; third, calculations; fourth, comparisons; and fifth, chances of victory.

Quantities derive from measurements, figures from quantities, comparisons from figures, and victory from comparisons.

Sun Tzu

With all the measurements and comparisons leading to concentrating at a decisive point, the following guidelines place the process in a viable perspective:

First, you cannot be all things to all people. Most huge, global organizations shy away from pursuing such a generally fruitless effort, unless they have sufficient resources to be a segment leader in all markets, which is not the typical case.

Second, where there is an overwhelming leader in the market, your best strategy is to concentrate your strength and strike one decisive point that represents your competitor's weakness.

Third, your best course of action is to acquire market intelligence that documents the specific needs and problems of a well-defined customer group and then concentrate your resources.

Fourth, look for a customer segment that represents a market trend, or find a niche that has been overlooked, neglected, or poorly served—as was the case with the boomers.

There is one additional guideline that deserves your attention: the ability to latch on to relationship marketing, which by extension becomes a decisive point in a comparative analysis. This means building long-term satisfying relations with key groups—customers, suppliers, distributors—to retain their long-term preference and business. The intent is to deliver high quality, distinctive service, at competitive prices to customers.

Relationships can run the gamut from a minor one where a salesperson makes a one-time transaction of selling a product, to a more integrated relationship where partnering means working consistently with a customer to discover ways to resolve problems and generate savings.

Or it can mean helping the customer design a product for *its* customers. In some instances relationships can include placing an individual on the customer's premises to assist in a variety of tasks from inventory control to providing technical assistance. Doing so in these examples may work to keep your firm locked in and the competitor locked out.

Implementing the transition from transaction selling to relationship marketing is handled in the following ways: First, identify the key customers warranting relationship marketing. Then train the salesperson or other contact individual to deal exclusively with the customer. In turn, that requires a customer-relationship plan that details the objectives, actions, and required resources to implement the program. Also make the Internet and related technology an integral part of the customer-relationship plan.

> Concentration seems easy, but is quite hard in practice. Everybody knows that the best way is to use a large force to defeat a small one, and yet many people fail to do so.
>
> Mao Tse-tung

IDENTIFY A DECISIVE POINT IN A MARKET SEGMENT

To assist you in identifying a decisive point in a segment, use the following criteria. Note, too, these guidelines would augment your comparative analysis.

Measurable. Can you quantify the segment? For example, you should be able to quantify how many factories, how many engineers, or how many people with or without access to technical service exist within the market segment.

Accessible. Do you have access to the market through a dedicated sales force, distributors/dealers, transportation, or the Internet?

Substantial. Is the segment of adequate size to warrant your attention as a viable segment? Further, is the segment declining, maturing, or growing?

Profitable. Does concentrating on the segment provide sufficient profitability to make it worthwhile? Use your organization's standard measurements for profitability, such as return on investment, gross margin, and profits.

Compatible with competition. To what extent do your major competitors have an interest in the segment? Is it of active interest or of negligible concern to your competitors?

Effectiveness. Do your people have acceptable skills and resources to serve the segment effectively?

Defendable. Does your firm have the capabilities to defend itself against the attack of a major competitor?

Answering those questions will help you decide on a market segment with good potential for concentrating your resources, as well as for gaining ample information about your customers and competitors. Once selected, the above criteria can be used to test the feasibility of a market segment. (While computer software may speed up the segmentation process, be certain the above criteria are included when you select the program.)

CLASSIC TECHNIQUES FOR SELECTING A MARKET SEGMENT

Table 3.2 displays the four most common ways to segment a market, based on demographic, geographic, psychographic, and product attribute factors. Each of these approaches can be subdivided into additional niches. Or they can be used in various combinations to create fresh opportunities. We now examine each.

Demographic Segmentation

Demographic variables are among the most widely used segmentation approaches. They owe their popularity to two facts: First, they are easier

TABLE 3.2

Bases for Market Segmentation

Demographic Segmentation	Psychographic Segmentation
Sex	Lifestyles
Age	Psychological variables:
Family life cycle	Personality
Race or ethnic group	Self-image
Education	Cultural influences
Income	
Occupation	
Family size	
Religion	
Home ownership	
Geographic Segmentation	**Product Attribute Segmentation**
Region	Usage rate
Urban/suburban/rural	Product benefits
Population density	
City size	
Climate	

to observe and/or measure than most other characteristics. Second, their breakdown of gender, age, family life cycle, race or ethnic group, education, income, occupation, family size, religion, and home ownership are often closely linked to differences in behavioral patterns.

In many instances, you can combine demographic variables to produce a more meaningful breakdown than just relying on a single criterion. For example, it is common to combine the age of the head of the household with the family size and the level of household income.

If four age levels, three family sizes, and three income levels are distinguished, a total of 36 segments result. Using a combination of primary data, secondary data, and judgment, you can then determine the value of each segment and thus arrive at a well-thought-out conclusion about which segments represent primary and secondary decisive points.

Watch out, however, for unrelated demographic characteristics that could be unreliable: Gender may produce marginal differences in the usage patterns of cell phones or in the consumption of toothpaste and soft drinks. Or chronological age is not always a reliable indicator of behavioral patterns. And income level may prove relevant only when used with other variables such as social class, family life cycle, and occupation.

Geographic Segmentation

Geographic segmentation is relatively easy to perform because the individual segments can be clearly defined on a map. It is a sensible strategy to employ when there are distinct differences in climatic conditions, access to transportation, proximity to round-the-clock service or repairs—as well as such geographic considerations as varying regional tastes or unique culture-based habits and behaviors.

Geographic segmentation even extends to facial features used in advertising. When *Kodak* originally introduced one of its cameras worldwide, the company quickly learned through adverse market feedback that potential consumers in many countries around the globe, from the Philippines to India, from Hong Kong to South Africa, could not relate to the white girl portrayed in the advertising. Kodak promptly modified its advertising by using local models, which contributed to a phenomenal success story.

Internationally, blocks or clusters of countries can often be approached in a similar fashion, particularly if they share the same language and cultural heritage. For instance, in Latin America the same advertising media are often appropriate for several countries.

While there are numerous cultural differences in many of those countries there are common problems with shared features, known as *cultural
universals*. These include economic issues, marriage and family rituals,
educational systems, religious observances, and supernatural beliefs.

Geographically, you can segment by region, city size, population density,
or other geopolitical criteria. However, such segmentation is effective only
if it reflects differences in need and buying patterns. Some entrepreneurs,
for example, can adjust their advertising efforts to target very narrowly
defined groups.

For instance, *Panera Bread Company*, a restaurant chain, has maintained
prices and in some cases increased them even during a recession-wracked
economy. Its management bucked conventional industry wisdom by
avoiding discounts. Instead, they targeted customers who can still spend
an average of about $8.50 for lunch, where other chains offered meals for
as little as $5.

Further, Panera has been able to persuade customers to pay premiums
because it has been improving the quality of its food. "Most of the world
seems to be focused on a segment of the population who are unemployed.
We've focused on the 90% that are still employed," stated a senior Panera
executive.

Psychographic Segmentation

This form of segmentation results from the application of psychographic
variables, such as lifestyle, personality, and self-image. Banks, food manufacturers, liquor producers, and restaurants benefit from the advantages of
psychographic segmentation.

You may be able to observe the behavioral nuances of distinct target
groups on your own. For instance, you might notice how some groups that
fit your customer profile live, spend, and behave. You can then explore
the activities they get involved with, interests they pursue, and opinions they hold. If you wish to go into further depth on behavior, you may
need the professional help of individuals trained in some aspects of the
social sciences.

The intent here is to open your thinking to the variety of ways to
look at customer groups, so that you can look for a decisive point when
segmenting a market. Therefore, it is possible for you to use a form
of lifestyle segmentation by observing the behavior of distinct target
groups.

Product Attributes

Product attributes include usage rates defined as nonuser, ex-user, potential user, first-time user, and regular user groups.

In practice, such information is further broken down to distinguish user groups as light, medium, and heavy users of the product. Often, heavy users of a product represent a relatively small share of total households or corporate buyers, yet account for the major portion of the sales volume in the market.

These breakdowns, in turn, would trigger different motivational appeals to improve the level of responsiveness from various segments. In other applications, companies with high market share might be especially eager to attract potential users, while smaller competitors with lower market share would devote their efforts to converting existing users.

Overall, you will find product attributes most practical in segmenting a market, and particularly applicable in deciding where to deploy a sales force, allocate budgets, prioritize product development projects, and direct promotional campaigns for a concentrated impact.

> Strategy, besides indicating the decisive points, requires two things: First, that the principal mass of the force be moved against fractions of the enemy's; second, that the best direction of movement be adopted, that is, one leading straight to the decisive points already known.
>
> **Jomini**

ADVANCED TECHNIQUES FOR SELECTING A SEGMENT'S DECISIVE POINT

Beyond the more conventional techniques listed above, there is an advanced procedure that uses a strategy-driven approach to selecting a market segment and concentrating at a decisive point. The technique consists of classifying markets as *natural, leading edge, key, linked, central, challenging, difficult,* and *encircled.* Using these eight categories as a guide, you can look with a more critical eye at what challenges you will face. Then, you will be better able to assess the risks and potential rewards when selecting your strategies. As you examine the characteristics for each market category, you may find some overlapping. That is acceptable as there are inherent commonalities among the various markets.

> Generally the best policy is to take a state intact; to ruin it is inferior to this.
>
> **Sun Tzu**

Natural Markets

In this type of marketplace, you operate in the familiar setting of your established markets. The implication is that within such customary surroundings, personnel tend to be at ease and may not be motivated to venture out of their comfort zone.

Yet, to expand, you have to motivate them to move beyond the confines of existing markets. That means you should get back to the issue of your organization's culture. That is, does your organization's culture permit venturing out of familiar territory?

For the most part, in a natural market, you and your rivals have learned to adopt a live-and-let-live policy—"to ruin it is inferior to this." That condition exists as long as each company sticks to its own dedicated segment. Generally, outright aggressive confrontations are seldom used.

The primary reason for this uncharacteristic display of togetherness in a highly competitive world is that you and your rivals share a common interest in furthering the long-term growth and prosperity of the market.

On the other hand, if any one competitor chooses to move forward and gain a meaningful benefit, here is where you have to decide on your goals. If your aim is to check the expansion of the competitor, then you have to look for your possible comparative advantage in Table 3.1. How aggressive you choose to be entails looking at your strategic viewpoint about maintaining stability in the marketplace.

There is one additional dimension that characterizes this category, which you should actively keep in the forefront of your thinking: Industries, markets, and products go through successive life cycle stages of introduction, growth, maturity, decline, and phase-out. Much of the movement through these stages is driven by the rate of adoption of technology, which can affect changes in consumer behavior.

There are also variations triggered by legislative issues and the current trends about the environment. Unless there is an industry-wide movement to deal with such concerns, these are generally beyond your ability to control. Therefore, your best course of action to sustain growth is to take the lead in searching for new niches in which to concentrate.

Leading-Edge Markets

Leading edge means exploring markets by making minor penetrations into a competitor's territory. The intent is to investigate the possibility of opening another revenue stream. Therefore, you want to acquire the following types of intelligence:

> The feasibility of the market to generate a revenue stream over the long term; and if possible to grow into additional niches.
>
> The amount of investment needed to enter and gain a foothold in the market and then expand to secure a profitable segment (see the section above, "Identify a Decisive Point in a Market Segment").
>
> A time frame for payback and eventual profitability.
>
> An assessment of competitors: their market positions, strengths/weaknesses, decisive points, and the nature of the opposition you will likely face.

A classic example of a leading-edge market is the initial penetration by a few Japanese companies into the North American market with small copier machines. *Xerox*, the market leader in those early years, concentrated its marketing efforts at large corporations with a line of large copiers.

Xerox managers initially avoided the small copier market. That oversight proved to be a critical error. Armed with a low-cost, no-frills desktop copier, enterprising Japanese copier makers found their decisive points and moved in virtually unopposed and exploited a wide-open opportunity in the vast market of small and mid-sized firms. Once established, they moved upscale in a segment-by-segment assault and took over a significant amount of Xerox's primary market share.

> The major battle must always be considered as the true center of gravity.
>
> **Clausewitz**

Key Markets

Key means that you and many of your competitors seem evenly matched. The general behavior is that you would not openly oppose an equally strong rival.

However, you may find that a competitor is attempting to dislodge you from a long-held position with the clear aim of taking away customers or disrupting your supply-chain relationships. Then you may be forced to launch a counter effort by concentrating as many resources as possible to blunt the effort. Such actions are appropriate, however, if they fit your overall strategic objectives.

Therefore, keep the big picture in mind: If you expend excessive resources in hawkish-style actions such as price wars, then you may be left with a restricted budget to defend your market position. In any event, you want to make sure that any confrontational action you take is the "true center of gravity."

Linked Markets

In this category, you and your competitors are linked with easy access to markets. Your best strategy is to construct barriers around those niches (decisive points) that you value most, and from which you can best defend your position.

Barriers you can create include

Above-average quality
Feature-loaded products
First-class customer service
Superior technical support
Competitive pricing
On-time delivery
Generous warranties
Patent protection

Not only do you build barriers against competitors' incursions, you also benefit by solidifying customer relationships. In particular, customer loyalty gives you a long-lasting, profit-generating advantage that is difficult for a competitor to overcome.

It is the one area that makes a meaningful addition to your growth. As one management analyst put it: "If you currently retain 70 percent of your customers and you start a program to improve that to 80 percent, you'll add an additional 10 percent to your growth rate."

If a mere demonstration is enough to cause the enemy to abandon his position, the objective has been achieved.

Clausewitz

Central Markets

Central means that you may face powerful forces that threaten your market position. These forces are as diverse as watching competitors eat away at your position through aggressive pricing, or by offering dazzling feature-laden products, or by technology-rich firms generating new applications overlaid with value-added services.

To counter such threats, look for joint ventures so that the cumulative effects yield greater market advantages and offer more strategy options than you can achieve independently. A "mere demonstration" in the form of a merger and acquisition (M&A) or other type of joint venture has proven the strategy of choice for some companies.

Challenging Markets

If you enter a market dominated by a strong and aggressive competitor, be watchful. You could place your company at excessively high risk.

If, however, your long-term objectives strongly support maintaining a presence in a challenging market, and if the expenditures of financial, material, and human resources are consistent with your overall strategy, then find a secure position, again, at the decisive point. It could be one of your single best chances for lessening the risk and achieving a solid measure of success.

Dell Computer is a prime example of employing excellent supply-chain management. From its beginnings, the company's strategy relied on activating its manufacturing process and supply-chain only when an order was received from a customer. That strategy worked at eliminating the cost of storing excessive inventory.

Dell benefited by shipping just the right amount of components to its factories, thereby avoiding investing in expensive warehousing. For instance, in one facility, what used to be done in more than two buildings now is accomplished in one by applying the techniques of supply-chain management.

A single center of gravity ... is the place where the decision should be reached; a victory at that point is in its fullest sense identical with the defense of the theater of operations.

Clausewitz

Difficult Markets

This type of market segment is characterized as one where progress is erratic and highly competitive. If attempting to make any meaningful market penetration, secure key accounts, or maintain reasonable levels of logistical support, you are likely to be blocked by asset-draining barriers.

Also, if a competitor is fully prepared, takes you off guard, and you subsequently lose your market position, it is difficult to return to your former position. In effect, you are entrapped in an untenable condition and your entire business strategy could be in jeopardy.

Your best course of action is to look for "a single center of gravity" and go forward, just as long as the effort is consistent with your mission and long-term strategic objectives.

The supreme excellence in war is to attack the enemy's plans.
What is of supreme importance is to attack the enemy's strategy.

Sun Tzu

Encircled Markets

Encircled segments foretell a potentially risky situation. That is, a market condition in which you control limited resources; and any aggressive action by a stronger, well-positioned competitor can force you to consider pulling out of a market.

Under those threatening conditions, your best approach is to maintain ongoing competitor intelligence. Your aim is to conduct a meaningful comparative analysis that would open opportunities by exposing your opponent's vulnerabilities so you can "attack the [competitor's] plans"—or single out a decisive point.

Doing so at least permits you to develop a contingency plan, which could include various options, such as developing a series of product enhancements on a phasing-in time schedule, launching value-added services that favor your strengths and highlight your rival's weaknesses, or preempting your competitor's promotional programs and thereby "attack the [competitor's] strategy" and his ability to maintain a profitable market position.

The essential aim, of course, is to attempt a turnaround and discourage your opponent from making a monumental effort to push you out. By taking a bold approach, you may even pull off a monumental coup by discouraging your competitor from pursuing an aggressive action against you. If your actions are well positioned, you may even place your competitor in his own encircled position!

The issue, therefore, is not which side has the greater resources. Rather, it is a case of who has the resolute mind-set that can outthink, outmaneuver, and outperform the competitor. Arming yourself with a process to conduct a comparative advantage (Table 3.1 and Table 3.2) offers a means to target a decisive point.

That said, keep in mind the following time-honored advice, which is totally applicable to today's business environment should a competitive situation turn negative.

> In war the result is never final ... merely a transitory evil.
>
> **Clausewitz**

Table 3.3 summarizes the eight categories of market segments.

> To overcome your enemy you must match your effort against his power of resistance, which can be explained as the product of two inseparable factors: the total means at his disposal and the strength of his will. The extent of the means at his disposal is a matter—through not exclusively—of figures, and should be measurable.
>
> But the strength of his will is much less easy to determine and can only be gauged approximately by the strength of the motive animating it.
>
> **Clausewitz**

Familiarizing yourself with the eight segment categories in Table 3.3 will equip you with additional insights when selecting segments. You will then be in a superior position to "match your effort against his power of resistance." Then, you can concentrate your resources rather than create an unequal distribution that dissipates your strength. Doing so therefore places you in an excellent position to exploit your opponent's weaknesses—and reduces his power to resist.

TABLE 3.3

Advanced Techniques for Selecting a Segment's Decisive Point

Category	Characteristics
Natural market	You and your rivals can operate harmoniously as long as each company sticks to its own dedicated segment. Aggressive confrontations are seldom used. All companies share a common interest in furthering the long-term growth and prosperity of the market.
Leading-edge market	Market entry means a minor penetration into a competitor's territory to determine the feasibility for generating a long-term revenue stream.
Key market	Competitors appear evenly matched within key market segments. The general strategy is that you would not openly oppose an equally strong rival. If the competitor attacks your position, then you are forced to launch a counter effort by concentrating as many resources as you can to "attack the enemy's plans."
Linked market	You and your competitors are linked with easy access to markets. Your best strategy is to construct strong barriers around those niches from which you can best defend your position.
Central market	You face powerful competitors that threaten your market position. Counter such threats by joint venturing with other firms. You thereby gain greater market advantages and strategy options than you can accomplish independently.
Challenging market	An aggressive competitor dominates the market and thereby could place your company at excessively high risk. If your long-term objective strongly supports maintaining a presence in the market, then find a secure position by locating your competitor's decisive point.
Difficult market	Competition is pervasive and market behavior is erratic. Gaining and maintaining market penetration is difficult. Overall, your best course of action is to go forward by conducting a comparative analysis and locating your areas of strengths and your competitor's weaknesses.
Encircled market	This market is risky where any aggressive action by a stronger, well-positioned competitor can force you out of the market. Maintain ongoing competitive intelligence to accurately assess the vulnerabilities of your opponent. If you lack a capability to mount a meaningful competitive response, then exiting the market is a prudent strategy.

As you develop your strategy of concentration at a decisive point, here are questions you should address—ideally, using a team approach:

Should you change the allocation of your resources after you have gained a favorable market position?

If you gained a dominant position, then you should move partially from the offensive to the defensive, making certain that you have an active defense against expansion-minded competitors.

You saw the effect of a somewhat lethargic response by the management of Xerox when Japanese copier companies first showed interest in the North American market and then moved energetically to concentrate in the small business segment as its initial point of entry.

Consequently, how you deploy your resources and, in particular, how you shape your marketing mix should change with the situation. That also means adding as much flexibility to your organization as feasible, which includes holding reserves. You thereby ready yourself to respond rapidly to counter competitors' moves.

Also, one of the essential activities related to properly allocating resources is to acquire as much competitor intelligence as possible. Armed with reliable information, you can budget funds for products that could challenge competitors' entries. What follows is that you are in an informed position to effectively deploy salespeople, motivate middlemen, and fund promotions.

The bottom line is that as you concentrate your resources, you thereby secure an advantage by forcing a competitor to spread his efforts and weaken his overall market position.

How will your personnel react under diverse market conditions— specifically, during actual competitive confrontations?

Unless you have observed your staff's behavior under strained situations, you are not likely to know the precise answer. Instead, it is safe to rely on the premise that only the skilled will survive competitive confrontations.

Therefore, as indicated earlier, the experience, training, and morale of your personnel are far superior to quantity. Consequently, do not sacrifice quality. If you do, there is a greater chance of failure—unless the quality of the competition's personnel is far inferior to yours.

Although no situation offers certain results, it is clear that those with the skills and high-level training to do so can turn shortcomings to an advantage. Conversely, those who lack skills are in jeopardy. The ultimate responsibility for the condition of your personnel remains with you.

As such, it deserves your highest priority and is an essential factor in leadership and managerial competence. Therefore, it should be your primary concern to monitor the actions of your personnel under a variety of circumstances.

Are you prepared if the competitive situation shifts from success to failure?
As an extension of the above, think about the behavior of your supervisory personnel. Could you count on them to resume control of a losing situation and regain the confidence of their staff should the business plan start to crumble? How are the senior and junior managers perceived by those they supervise? Is there an adequate level of trust and confidence in their ability to make effective decisions?

Consider times where rapid decisions are needed to alter a threatening situation. Are communication channels intact? Do they lead to understanding or confusion? Are there undercurrents of unspoken or disguised messages filtering through to the group that generate encouragement or discouragement, elation or fear?

How do you assess your personal performance?
Objectivity is the issue here. All too often, when gauging personal performance, there is a tendency to place blame solely on others. Yet, it is not the underlings who are always at fault. Rather, managers must take their share of blame in such areas as incompetent leadership, poor or nonexistent market and competitive intelligence, inadequate or unclear communications, ignorance of fundamental strategy rules, and an inability to motivate employees to implement plans with speed and enthusiasm. Under these circumstances, truth and impartiality would be the prudent paths to correct lagging performance.

> For to win one hundred victories in one hundred battles is not the acme of skill. To subdue the enemy without fighting is the acme of skill. Thus, what is of supreme importance is to attack the enemy's strategy.
>
> **Sun Tzu**

In summary, use the following guidelines to concentrate at a decisive point:

Make use of competitive analysis (Tables 3.1 and 3.2) to identify your competitor's weaknesses or market gaps.

Search out unserved, poorly served, or emerging market segments that represent growth and permit you to establish a foothold.

Secure your primary segment by using such strategies as strengthening your own brand, developing product enhancements, adding specialized applications, offering creative packaging, and offering value-added services. You thereby create a barrier by blocking a competitor's entry or by using them as part of an offensive strategy to expand.

Recognize that every market segment you enter is actually the starting point of a new operation and a new sales cycle, which opens up a whole set of fresh possibilities.

No campaign is complete without first determining where to concentrate your efforts. Your aim is to apply maximum effort at a decisive point. That point means positioning your resources to satisfy customer demands, concurrent with preempting your competitor from taking similar action against you.

Consequently, a dominant rule of strategy is to keep one's forces concentrated: first in general and then at the decisive point. The immediate application is that a numerically inferior organization can win by concentrating resources at a decisive point. Yet to accomplish this feat takes outstanding leadership and strategy skills by employing boldness, indirect approach, surprise, and competitive intelligence.

To the extent that you can, the aim is to avoid costly confrontations that drain resources and often accomplish little. Rather, "to subdue the enemy without fighting is the acme of skill."

The greatest talent of a general, and the surest hope of success, lies in some degree in the good choice of these points—objective points and decisive strategic points.

Jomini

4

Create a Lifeline to Business Strategy: Employ Competitor Intelligence

> Know the enemy* and know yourself; in a hundred battles you will never be in peril.
>
> When you are ignorant of the enemy, but know yourself, your chances of winning or losing are equal.
>
> If ignorant both of your enemy and of yourself, you are certain in every battle to be in peril.
>
> **Sun Tzu**

Chapter 3 placed strong emphasis on acquiring market and competitor intelligence when searching for a comparative advantage and for locating a decisive point on which to concentrate. Similar advice was stressed in Chapter 1, in the discussion on maneuvering by indirect strategy.

As an extension of these topics, and within the framework of Sun Tzu's sage comment that "you will never be in peril" by knowing your competitor and knowing yourself, the central aims of this chapter are to provide you with compelling reasons to, first, raise the priority level for acquiring competitor intelligence, and second, integrate the findings into developing your competitive strategies.

Consequently, given the scope of information required to develop competent strategies, it is up to you to look at a competitor with a 360-degree view. That is, use a circular approach to open your eyes to a changing marketplace of evolving technologies, environmental trends, and shifting buying patterns among your customers.

In this panoramic scene, seek information and insight that can impact your decisions about selecting markets, launching new products, and devising

---

* As in previous quotes, substitute *enemy* with its business equivalent, *competitor* or *rival*.

competitive advantages. You will improve your chances for establishing a formidable defense in those segments that represent your core business.

In all, reliable intelligence helps you outthink, outmaneuver, and outperform your competitor before he can react to *your* movements. It also helps you assess the level of risk associated with going after an opportunity.

> All warfare is based on deception.
>
> **Sun Tzu**

The intelligence-gathering approach begins by looking at a particular category of competitor intelligence that feeds three foundation components of strategy: (1) indirect strategy, (2) concentration at a decisive point, and (3) surprise. That category is *deception*. It is one of the earliest and most basic parts of the intelligence-gathering process. Sun Tzu and other sages considered it a key factor for success long before the beginnings of organized intelligence.

For business application, the purpose of deception is to create a diversion or distraction; in that way you reduce a competitor's resistance against your efforts.

The essential points: If you know your rival's plans and are able to monitor his moves, then you can estimate with some degree of accuracy which strategies are likely to succeed and reject those with minimal chances of success. You are then in a better position to concentrate your efforts at a competitor's decisive point.

You also improve your chances of establishing a formidable defense in the segments that represent your core business. In all, it helps you outthink, outmaneuver, and outperform your competitor before he can react in time to interfere with your movements.

Creating a diversion that borders on legal deception may even cause discomfort for some individuals. Certainly, if deception is carried to extremes, it can potentially impact some aspect of corporate ethics and even lead to legal issues.

> The most successful soldier will always make his attack not so much by open and regular approaches as by ... deceiving our enemies.
>
> **Thucydides**

Today, "deceiving our enemies" takes many forms: hackers break into government and corporate databases with regularity, offices are bugged, employees steal secrets and sell them to any competitor for pure monetary gain, managers pass on confidential information about new products through surreptitious agreements with a rival company. In addition, there are foreign governments that sanction cybercrimes by stealing intellectual property as a part of their national policy. The most blatant illegal examples include the following:

Apple Computer: A global supply manager at Apple was caught sending drawings, pricing information, and projected sales figures on various Apple products to suppliers in China, South Korea, Taiwan, and Singapore. Suppliers paid kickbacks totaling more than $1 million to the individual.

Dow Chemical: A research scientist who worked at Dow for 27 years stole trade secrets for making chlorinated polyethylene and sold them to Chinese companies.

DuPont: A research chemist working on organic light-emitting technology secretly downloaded files to an external drive. He then covertly took a job with Peking University, which was working on the same technology. He went on to make personal presentations about commercializing the research.

Ford Motor Company: A product engineer, before leaving Ford, copied 4,000 documents, including design specs. As he was about to leave the United States on a one-way ticket, the data turned up on his laptop when it was confiscated at an airport.

Goodyear Tire & Rubber: Two engineers at a Goodyear supplier were convicted of entering a Goodyear plant, photographing a device used to wrap cable for a tire's inner thread, and transmitting the photos to a rival company.

Motorola: An engineer was charged with economic espionage for stealing trade secrets to benefit a foreign government. The individual was caught at a U.S. airport with 1,000 Motorola documents and a one-way ticket to China. Among the papers were confidential documents relating to Motorola's mobile-phone technology.

Boeing/Rockwell: An engineer was caught with 300,000 pages at his home, including technical information on a U.S. space shuttle, F-15 fighter, B-52 bomber, and Chinook helicopter.

Other victims of espionage include *Google, Lockheed Martin, Chicago Mercantile Exchange, American Superconductor,* and *Dow AgroSciences.*

> I make the enemy see my strengths as weaknesses and my weaknesses as strengths.
>
> **Sun Tzu**

What implications do these cases of deception have for competitive strategy? From a macro viewpoint, stealing a company's technology, seizing advance research, and accessing any other area that converts data to a product or service advantage can seriously jeopardize a manager's strategy options and the ability to concentrate at a decisive point. At the micro level, such outrageous acts explicitly place a manager in a high-risk situation by blocking him or her from building strong defensive barriers to protect an existing market position.

From your viewpoint, similar situations can realistically gut your efforts for developing expansion strategies by stripping away at the uniqueness of your product—or other parts of the marketing mix.

Deception is a two-pronged issue: First, maintain security and guard against rivals stealing your competitive advantage and turning it around to their benefit and against you. Second, within legal and ethical bounds, use deception as a way to maneuver by creating surprise, finding the decisive point of concentration, and developing an indirect strategy.

In the context of deception, therefore, where secrets are embezzled by individuals who are driven by assorted motivations, it is appropriate to discuss the *people* part of intelligence gathering. In turn, that leads to the use of people as *agents* for acquiring and disseminating information.

Engaging in such stealth activities is usually contrary to the type of practices most managers care to undertake. Yet, as pointed out above, within ethical and legal bounds and in a highly competitive world, deceptions for diversion and distraction purposes are legitimate weapons. And employing agents is an essential component of the competitive intelligence process.

Agents serve a highly credible role in interpreting industry events, organizing stray pieces of competitive information into coherent intelligence, and flushing out information about a competitor's plans. They also contribute to disseminating deceptive information so that the competitor "sees my strengths as weaknesses and my weaknesses as strengths." Consequently, it is in your best interest to use the following guidelines when employing agents.

> Secret operations are essential; upon them the army relies to make its every move. An army without secret agents is exactly like a man without eyes or ears.
>
> **Sun Tzu**

EMPLOYING AGENTS

Agents are your "eyes and ears" at conferences, trade shows, and even at your competitor's locations. They go beyond the analytics, charts, surveys, benchmarking, and other intelligence-gathering techniques.

They explore the human side of competitor intelligence by reporting on the behaviors and personalities of key individuals. Their primary tools are personal interaction and observation. Agents also screen and interpret events, news, and validate or dismiss information gathered by other means.

Before moving forward and employing agents, however, observe a few general cautions: First and foremost, make certain that you are not violating ethical and legal guidelines, as cited in the above examples. Check, too, that your use of agents agrees with your company's policies. Second, assess prospective agents' motivations, personality traits, and talents. Then, you can determine in what capacity to employ them. For instance, some individuals' primary pursuit is money, and they have a minor interest in obtaining accurate information about the competitor. In such cases, question their integrity and take great care in using them. Third, develop a clear idea about the information you seek. Then make certain they understand what you want.

> Among agents there are some whose only interest is in acquiring wealth without obtaining the true situation of the enemy, and only meet my requirements with empty words. In such a case I must be deep and subtle.
>
> **Sun Tzu**

The following represents categories of agents, along with suggestions on how to use them.

Native Agents

Native agents are the types of individuals with whom you would normally interact during professional gatherings. They tend to volunteer company information to satisfy their personal interests, make new contacts, and advance their careers. Often, they are somewhat uncaring about their respective company's security. Or they were not cautioned about the dangers of revealing company secrets. You will find native agents in a variety of places. Trade shows serve as fertile venues for gathering intelligence from those individuals. It is also a place where competitors typically reveal extensive information through elaborate demonstrations about their products and then freely distribute literature overflowing with facts about pricing, backup services, logistics, and product specifications. Also, key executives from competitors' organizations often present papers at open meetings, which detail sensitive information about upcoming products, services, and even market-entry plans. Then there is the Q&A period where the speaker, trying to further impress an audience, impulsively pours out classified information.

Another prime area for intelligence gathering is the familiar hospitality suite at trade shows and professional meetings where alcohol and talk flow freely. It is a site where security is often lax and everyone's guard is down.

College classrooms represent another source of information, particularly where part-time instructors are also executives and use their respective companies as case examples. In reverse, students often reveal data about their companies through class presentations, term papers, and casual conversations.

> The first essential is to estimate the character of the spy to determine if he is sincere, truthful, and really intelligent. Afterwards, he can be employed.
>
> **Sun Tzu**

Inside Agents

As with native agents, inside agents work for competitors. In many cases, they may have been bypassed for promotion, feel underpaid and underappreciated, relegated to an insignificant job, or generally pushed aside in a variety of political or power struggles within the organization. They feel abused and see their careers languishing unless they make some bold move. They may also find themselves surrendering to financial pressures to keep family and self whole. And their attitude may be now-or-never. You need to assess such individuals carefully for their stability and determine how to use them judiciously. Obviously, you want their information, within the bounds of ethical and legal guidelines.

Beyond personal observations, you would employ inside agents for their expertise to sort out meaningful information from scientific and professional journals, industry studies, or from innovative projects described in articles and professional papers written by the competitors' employees. Product literature and product specification sheets readily available at trade shows and meetings are packed with tremendous detail. Your agents should be able to interpret the data for meaningful intelligence.

In-house company newsletters and news releases contain a fountain of information about individuals who have left a competitor's employment and may have moved to the consulting circuit. If approached, these former employees may be willing to reveal information—unless specific contractual restrictions apply.

Press releases may include new employee announcements along with job descriptions, contracts and awards received, training programs available, office or factory openings or closures. You may also find specific news items that reveal competitor's current plans. Here, again, inside agents could handily provide useful analyses.

Beyond the above listing, there is the continuing flow of rumors from customers and suppliers that your agents can sort out and verify. Additionally, there are local sources worth tapping, such as banks, local trucking companies, and real estate offices.

> Knowledge must come from the double agents, and therefore it is mandatory that they be treated with the utmost liberality.
>
> **Sun Tzu**

Double Agents

These agents try to extract intelligence about your company. Stay alert to their intentions. Once identified, you can attempt to turn them around and get them working on your behalf. If you are successful, they could serve as your inside agents. Here, too, you can assume double agents seek lavish rewards and may even show similar personality traits and motivations to inside agents. However, it is in your best interest to exercise caution. That is, determine the sincerity of these individuals, the reliability of their information, and how long you can expect them to remain loyal to your cause. Once again, make certain you are not violating ethical, legal, or policy guidelines.

Expendable Agents

These agents are your own people who are deliberately fed inaccurate information, which is disseminated in a variety of ways to distract competitors into making wrong decisions. Such contrived leaks take many forms: Passing fabricated information about new product features through sales reps who come in contact with competitors' reps. Or it could be product managers revealing false dates about a product launch that would disrupt a competitor's plans.

In spite of your possible discomfort when undertaking such activities, look at the situation from strictly a strategist's viewpoint. Disinformation for the purpose of deception needs distribution to divert competitors from directly opposing your moves and consuming resources that you cannot replace.

You thereby preserve your company's hard-won market position, control needless expenditures of financial and human resources fighting unnecessary market battles, and avoid disrupting your strategies.

> Of all rewards none is more liberal than those given to secret agents.
>
> **Sun Tzu**

Living Agents

These agents usually provide the most credible information. They are generally experienced, talented, and loyal individuals who can gain access to, and become intimate with, a competitor's high-level executives. They sit in a position to learn their plans and observe movements. These individuals are truly the eyes and ears and often enjoy the closest and most confidential relationships.

> And therefore only the enlightened sovereign and the worthy general who are able to use the most intelligent people as agents are certain to achieve great things.
>
> **Sun Tzu**

Perhaps the one unsettling issue to cope with when using agents is which one of your employees, knowingly or inadvertently, passes on your company's sensitive information to competitors. Where possible, make every effort to find out the perpetrators and obtain clues about

TABLE 4.1

Questions You Should Ask of Agents

What are the competitor's overall business strategies, particularly ones that would impact my plans?

What is the competitor's financial picture, including breakdowns of costs and sales by product lines?

What new products or services are under development?

What new markets are expected to be targeted?

How are the competitor's business units staffed and organized, especially in marketing and product development?

What is the caliber of the competitor's leadership?

What market positions or market shares do key competitors hold within each product segment? Are there plans to increase, maintain, or reduce their respective positions?

Where are the competitor's vulnerabilities, which could represent a decisive point for a concentrated effort, e.g., product depth, product quality, customer service, price, distribution, marketing, and reputation?

what motivated their behavior. It may be comforting to know that many of those individuals are often exposed—ideally, before too much damage has been done.

Table 4.1 provides general questions for guiding you "to use the most intelligent people as agents." You may wish to add your own specific questions for agents to probe that would further pinpoint a unique strategy advantage.

> Now the reason the … wise general conquers the enemy whenever they move and their achievements surpass those of ordinary men, is foreknowledge.
>
> **Sun Tzu**

MARKET SIGNALS

Within the totality of competitor intelligence is "foreknowledge," which is composed of factual information and the more elusive area of market signals. It is in the competitive marketplace that pulsates with competitive actions where you can tune in to market signals. Here, you gain invaluable insights to enhance the accuracy of your decision making. It is also an area where qualified agents can offer valuable interpretations to feed the strategy process.

For instance, there are particular jarring signals that should alert you: Competitors open or close regional offices or plants. Sudden management changes are announced. New layoffs are revealed. Or upsetting rumors persist of new competitive alliances.

Not to be overlooked are the telltale signs of internal disorder and inept leadership as you observe competitors' personnel. For instance, managers openly show discouragement, display low morale, or exhibit short tempers. Sales reps overtly look for other jobs, criticize working conditions, or complain about shortages of sales aids and supplies. They whisper about ineffectual leadership and cuts in wages. They whine about stringent rules restricting travel-related expenses. Or they object to executives excusing (or overlooking) abuses in corporate procedures.

At the same time, you may observe signs of general disorder, sloppiness, or indications of internal desperation. For instance, there may be signals that represent changes in a competitor's traditional operating style, or in patterns of handling customer relationships, or in the general demeanor of executives as they interact with their personnel.

Other signals may come from customers who openly complain about the competitor's policies, rules, and procedures. If pumped for detailed information, they often come forward with a flood of grievances. All these are signals of fear, uncertainty, insecurity, and a variety of deep-seated internal problems. It is in such a state of unbalance and discontinuity that you can spot opportunities.

Consider, too, the possible implications of the following market signals on your operation:

A competitor abruptly announces a new value-added service. As important, the news triggers sudden interest among your customers.

A competitor introduces generous financial incentives for distributors to aggressively push their products. And your customers show strong signs of responding to those incentives.

Unforeseen promotional bursts from competitors siphon off sales you counted on.

A competitor suddenly shifts sales and service reps to a market segment that you thought was secure.

An enhanced product quietly and abruptly introduced by a competitor to your customers suddenly stirs interest.

An organizational shake-up at one of the key competitors indicates a new leadership team is taking charge.

The significant point is that it is in your best interest to maintain open channels of communications from a variety of sources to flag significant market signals for priority handling.

> It is sufficient to estimate the enemy situation correctly and to concentrate your strength to capture him. There is no more to it than this.
> He who lacks foresight and underestimates his enemy will surely be captured by him.
>
> **Sun Tzu**

TOOLS AND TECHNIQUES OF COMPETITIVE INTELLIGENCE

In addition to the issues discussed above, there are a variety of conventional tools and techniques "to estimate the [competitor's] situation correctly and to concentrate your strength." The following list itemizes the primary techniques. Select the ones based on the time and resources available to you. In any case, make every effort to use as many techniques as possible.

Sales Force

If you can, work with a sales force, either your own or a distributor's reps, to gather relevant data from those front-line individuals. Talk with them. Travel with them. They are in continuing contact with customers and are prime sources of real-time intelligence.

To maximize their output, instruct them on what to observe and brief them on the reasons for doing so, which may involve sharing some information about your business plans. Also show them why competitor intelligence is the critical underpinning of all business strategy and a prime source of fresh opportunities.

A highly useful approach is to require call reports after all customer visits. While product usage, future sales prospects, and similar information are the usual topics reported, get them involved in adding specific facts about potential market opportunities.

Especially emphasize—to the point of requiring them—reports on strengths and weaknesses of competitors, any significant changes in their strategies, and all other relevant activities. Be certain you read and react to their reports. The last thing sales reps want to hear is that their information is not being read or even acknowledged.

Customer Surveys

If your budget allows, use an outside market research organization to track competitors' activities and test product and marketing concepts through focus groups. The investment could have a big payoff for you.

If you cannot use a professional organization, try some of the following surveying techniques on your own:

In-person interviews: Question individuals with the aim of gaining insights about competitors. Such questioning could reveal gaps that can be converted into opportunities. Those interviewed could be users of your product or service, competitors' sales reps, and distributors.

Telephone interviews: These interviews are somewhat difficult to secure with the prevalent use of voice mail and a variety of electronic devices. However, if handled by individuals who can gain the interviewees' attention because of their reputation, or by trading hard-to-find information, then this technique could be a good source of information.

Surveys: Questionnaires by mail or Internet are useful to acquire product information about competitors' products and services. In such instances, you can avoid using your company name. Instead, use the name of a generic-sounding research organization. In such research, and depending on the type of product, it is useful to offer an incentive for taking the time to fill out the questionnaire.

Social networking: Increasing attention is being given to listening in on the online conversations on Facebook, Twitter, and YouTube about products, services, and companies. *Procter & Gamble*, for instance, launched social media campaigns for its big brands. It targeted 25- to 54-year-old women who are above-average Facebook users, and were the point of P&G's concentration.

Even the ultimate outcome of a *(conflict)* is not always to be regarded as final. The defeated state often considers the outcome merely as a transitory evil, for which a remedy may still be found … at some later date.

Clausewitz

In one instance, a survey conducted by a publisher of trade magazines found to its surprise that there was very little overlap between the magazine subscribers, trade show attendees, and website users. Discovering that audiences were significantly larger than indicated by the subscriber lists opened up a high-value opportunity by which "a remedy may still be found."

Published Data

Look at the growing number of electronic databases that provide voluminous quantities of competitive information. Some of these databases are free and available through public libraries; others are available at affordable costs.

Tap into the numerous websites for valuable information. Even small-town newspapers are rich with information, such as a competitor's minor activities making front-page headlines. Other sources include trade journals that provide financial and product information about competitors. And want ads in print and over the Internet offer clues to the types of personnel and skills being sought.

If your competitors are public corporations, refer to their easy-to-access and information-crammed annual reports. With great detail, they reveal financial data, product development plans, and markets they view for future growth.

As indicated earlier, speeches by senior executives of competing companies offer valuable insights into their firms' future plans, industry trends, and strategies. At times it astonishes how much sensitive information is provided in speeches that are given at trade shows and professional meetings.

Government Agencies

Depending on your location, you can access an ample supply of information from government sources through direct access, service organizations, or over the Internet. These sources usually include a vast amount of data from national, regional, and local government offices. Most information is available simply for the asking, or at modest cost.

In some instances, information is available about types of work contracts previously awarded and to whom. Also, competitive information may be obtainable about which companies are currently bidding for new contracts, which provides valuable clues to the type of competitors eligible to bid.

Industry Studies

Management consultancies, universities, trade associations, individual consultants, and investment securities firms conduct industry studies. Also, specialized companies publish industry and competitive information, such as

- Data on a vast variety of products sold through retail outlets, television, magazines, and online
- Weekly family purchases of consumer products and information on home food consumption broken down by geography and demographics
- Reports of warehouse withdrawals to food stores in selected market areas
- Statistics on television and radio markets according to demographics, experiential lifestyles, and brand preferences

On-Site Observations

You can handle this form of competitive intelligence personally. For instance, simply observing numbers of service vehicles and supply trucks in and out of a plant or at store locations provides first-hand information about the level of activity at a competitor's location.

In one case, an observer noticed several freight cars at a competitor's railroad siding. However, on closer examination, there was a barely visible layer of rust on the tracks, which indicated little or no recent rail activity. The observer concluded that the freight cars were there as a deliberate deception.

Also observing the schedules of employees arriving and leaving work can indicate patterns from which you can make reliable assessments. Even watching the movements of individuals in a parking area, the body language of people talking, or the work materials being carried can give clues that further fit an overall pattern. As for a retail store environment, watching and recording individuals entering and exiting, with or without packages, could provide valuable information when compared to similar locations.

> One of the surest ways of forming good combinations … would be to order movements only after obtaining perfect information of the enemy's proceedings. In fact, how can any man say what he should do, if he is ignorant what his adversary is about?
>
> **Jomini**

Competitor Benchmarking

Benchmarking is another assessment tool to compare and measure your firm's business processes against those of your competitors; or as expressed by Jomini, "how can any man say what he should do, if he is ignorant what his adversary is about?" Therefore, use benchmarking to examine your business strategies against those of your competitors.

The value of this tool is that you learn how and why some competitors perform with greater success than other organizations inside and outside your industry. You will find such information highly useful when planning strategies and determining which actions are likely to succeed. *Xerox Corporation* is an outstanding example of successful benchmarking. The company developed the following 12-step process for competitor benchmarking:

Planning
1. Identify benchmark outputs.
2. Identify best competitors.
3. Determine data collection methods.

Analysis
4. Determine current competitive gaps.
5. Project future performance levels.
6. Develop functional action plans.

Implementation
7. Establish functional goals.
8. Implement specific actions.
9. Monitor results and report progress.
10. Recalibrate benchmarks.
11. Obtain leadership position.
12. Integrate processes fully in business practice.

Table 4.2 summarizes the benefits of benchmarking.

TABLE 4.2

Benefits of Competitor Benchmarking

Improves your understanding of customers' needs and sensitizes you to the underlying competitive dynamics operating within your industry

Helps validate the most effective strategies against those of your competitors

Helps you document why some competitors can perform similar processes at higher performance levels than at your organization

Creates a sense of urgency to develop long-term objectives and strategies

Encourages a spirit of competitiveness so that personnel recognize that performance levels in best-in-class organizations may exceed their own perceptions of what constitutes industry—and world-class—competitiveness

Motivates individuals to new heights of innovative thinking and achievement

---

> When capable, feign incapacity, when active, inactivity. When near, make it appear that you are far away; when for away, that you are near.
>
> Offer bait to lure him; feign disorder and strike him. Pretend inferiority and encourage his arrogance.
>
> **Sun Tzu**

To summarize, acquiring competitive intelligence uses a variety of approaches, including the use of agents; conventional tools, such as the Internet, databases, analytics, monitoring behavior, and observation; and benchmarking. These approaches are the prime ingredients to create deception, which in turn are the underpinnings for shaping indirect strategies, creating surprise, and determining the decisive point for concentration.

Further, by defining your competitor's operating patterns, you are more likely to determine his market position. You can then deploy your personnel and identify areas where you are likely to face the most or least opposition. Armed with such knowledge, you have lead time to take counteractions. What better way is there to establish personal credibility for your managerial skills than being able to motivate your staff to handle potential threats and opportunities?

Therefore, competitor intelligence as an information-gathering, decision-making tool that affects all operating parts of your business either directly or indirectly. It is the centerpiece of all offensive and defensive actions. By exposing strengths and weaknesses in your situation, as well as in those of your competitors, it functions as the core component when developing an indirect strategy and locating a decisive point for market entry.

If utilized with the same care you would give to a delicate instrument, competitor intelligence can signal subtle changes in the marketplace. For instance, it can help you preempt and outmaneuver competitors, preserve financial expenditures for exploiting the best opportunities, and even protect your firm's secrets from inquisitive onlookers.

> There is still another factor that can bring ... action to a standstill: imperfect knowledge of the situation.
>
> **Clausewitz**

Therefore, listen to the incoming intelligence and be sure to evaluate the reliability of each source. Then, you can validate or disprove reports. In either case you gain confidence and become more adept at making accurate decisions.

After you have assembled reliable intelligence and carefully deliberated on all that is meaningful, develop and implement your strategy with speed. It would be a far greater error to wait for a situation to clear up entirely. The reality of working in a dynamic competitive marketplace is that decisions are required—even in the fog of uncertainty.

Therefore, feel assured that the competitor intelligence you have assembled and screened, and the assessments you have made, will produce the results you expect. And even in the penetrating light of reality, should you discover that some intelligence is contradictory, false, or contains "imperfect knowledge," then you have little recourse but to lean heavily on your judgment and move on.

That means: Depend on your knowledge of the industry, trust in your years of experience, show confidence in those key individuals with whom you work, recognize the value of your formal and informal training, and rely on the richness of your intuition.

Taking it a step further, should pangs of doubt still persist, yet intuitively you know that rapid action is called for, if you are to prevent a potentially bad situation from deteriorating further, tell yourself that nothing is accomplished without shifting to the offensive. The only reason to pause or even retreat from a situation is to make preparations to eventually move forward.

Also, take comfort in this reality: The risk is quite low that your best-thought-out decision would easily ruin your company, providing you have gathered reliable competitor intelligence, and followed the rules of indirect strategy, speed, and concentration. And, if all your planning is kept within a mask of secrecy, the competition will not see where you are going.

Thus, what is of supreme importance is to attack the enemy's strategy.

Sun Tzu

Finally, the key points of this chapter are summarized in Tables 4.3 and 4.4. Table 4.3 describes the competitive intelligence process. Table 4.4 summarizes Sun Tzu's views on the roles of intelligence, surprise, and deception.

By intelligence we mean every sort of information about the enemy and his country—the basis, in short, of our own plans and operations.

Clausewitz

TABLE 4.3

The Competitor Intelligence Six-Step Process

1. Competitors' size	Categorize by market share, growth rate, profitability, as well as any other quantitative measures meaningful to your company and industry.
2. Competitors' objectives	Determine competitors' intentions related to product innovation, market leadership, global reach, regional distribution, and similar areas that would indicate a strategic direction.
3. Competitors' strategies	Analyze their internal strategies (speed of product development, manufacturing capabilities, delivery, and marketing expertise) and their external strategies (supply-chain network, field support, market coverage, and aggressiveness in defending or building market share.)
4. Competitors' organization	Examine organizational design, culture, operating systems, internal communications, leadership, and overall caliber of personnel.
5. Competitors' cost structure	Check on how efficiently they can compete, how long they can sustain pricing competition, the cost or difficulty of exiting a market, and their views about short-term versus long-term profitability.
6. Competitors' strengths and weaknesses	Identify decisive areas vulnerable to a concentrated effort, as well as those strong areas that should be avoided.

TABLE 4.4

Sun Tzu on the Roles of Intelligence, Surprise, and Deception

Attitude toward intelligence	Positive; optimistic
	Reliable intelligence can be obtained and is a major key to success
	Very useful
Main sources of intelligence	Spies (agents) and observers
The possibility of making rational decisions and forecasts	Rational, carefully calculated plans can be made on the basis of reliable intelligence
	Forecasting is possible; and careful planning is an important key to victory
Deception and surprise	Deception is the basis for all successful operations
	It is the weapon of choice
	Surprise can be achieved and is a key to success
The value of intelligence as a key to success	Make the utmost effort to obtain reliable intelligence
	Base all planning on intelligence, and make extensive use of deception
	Conversely, make the greatest possible effort to deny intelligence to the enemy
Problems	Excessive reliance on intelligence and deception, which can become a panacea
	Friction is underestimated, and the value of plans overestimated

5

Maintain High Performance:
Align Competitive Strategy with
Your Company's Culture

> Now, the supreme requirements of generalship are a clear perception, the harmony of his host, a profound strategy coupled with far-reaching plans, an understanding of the seasons, and an ability to examine the human factors.

> **Sun Tzu**

The "harmony of his host" and "human factors" that Sun Tzu refers to are intrinsic qualities of an organization's culture. They are the underpinnings that make the organization tick; qualities that put life into individuals' actions. As such, corporate culture is defined by its deep-rooted traditions, morals, values, beliefs, and history that power the organization.

As the operating system and nerve center of your organization, culture forms the backbone of your business strategy. It shapes how your employees think and react when entangled in a variety of internal and external situations—from organizational shake-ups to aggressive rivals attacking your market position. Thus, corporate culture operates in a dynamic environment, one that is never static.

> By moral influence I mean that which causes the people to be in harmony with their leaders, so that they will accompany them in life and unto death without fear of mortal peril.

> **Sun Tzu**

Consequently, to be "in harmony with their leaders ... without fear of mortal peril" is essential (and admittedly hard to achieve) if you are to maintain a positive, forward-looking operating culture. In particular, that means maintaining believable and supportive relationships with your staff.

When CEO Thorsten Heins took over at *Research In Motion* (RIM, maker of the BlackBerry™), he recognized the need for harmony during his high-stakes competitive encounters against the makers of iPhone® and Android devices. "The first thing you need to do with your employees is you've got to be open," he declared. "Tell them where we are. They're thinking it themselves anyways. But also, you have to get them excited about the future of BlackBerry. I don't like to be dictatorial, but if I have to, I can be. Right now at RIM, I'm a bit dictatorial."

The essential point is that there is no generically good culture for all seasons, just as there is no one-size-fits-all strategy for all competitive confrontations. However, you can count on universal attributes, principles, and concepts that are anchored to a historical platform that are proven to sustain a competitively healthy corporate culture. Table 5.1 summarizes the primary ones.

In contrast, there are traits that characterize a sluggish and lethargic corporate culture. Such an organization closes its eyes to global competition, vacillates over the competitive impact of new technologies, focuses only on building market share in existing markets as it avoids pushing the boundaries of new markets, and is generally passive to evolving trends.

Other negative signs include a conscious disregard for making the corporate culture compatible with an erratic marketplace, with the result that the organization loses its edge, complacency spreads, customer focus declines, and originality dries up.

> The contest of strength is not only a contest of military and economic power, but also a contest of human power and morale.
>
> **Mao Tse-tung**

To operate successfully in a vibrant corporate culture means tuning in to the "human power" of your firm's basic beliefs, moral codes, traditions, and standards of behavior. This approach will work for you as long as the foundations of your organization's or business unit's culture harmonize with the current competitive environment and your firm's overall strategic goals.

Such harmony is illustrated by *Netflix's* corporate culture, which is described as the *freedom and responsibility culture*. Management's aim is to

TABLE 5.1

Qualities of High-Performing Corporate Cultures

Beliefs and Morals	Employee Treatment and Expectations	Organizational Structure	Leadership and Business Strategy	Vision and Managerial Competence
Total commitment to customer satisfaction	Fairness in enforcing discipline	Use of cross-functional teams to cultivate originality and innovation	Managers show skill in developing and implementing competitive strategies and tactics	Expertise in business planning
Openness to new ideas	Independence with an entrepreneurial outlook	Rapid internal communications through a flat organization	Maximum use of competitive intelligence	Vision to locate new and evolving market opportunities
Tolerance for employee diversity	Flexibility in encouraging personal growth	Speed of decision-making	Growth strategy aligned to corporate culture	Ability to nurture the health and vibrancy of the organization
Respect for individual achievement	Continuous learning	Mutually supportive internal relationships	Sensitivity to market changes	Aptitude to conceptualize and communicate a strategic direction
Ethical conduct	Motivated employees	Authority and responsibility at the field level	Focus on customer retention	Facility to convert vision to action

(Continued)

TABLE 5.1 (Continued)

Qualities of High-Performing Corporate Cultures

Beliefs and Morals	Employee Treatment and Expectations	Organizational Structure	Leadership and Business Strategy	Vision and Managerial Competence
Personal fulfillment	High morale and pride in the organization	Innovative products and services based on a blend of innovation and technology	Emphasis on high morale through competent leadership	Sensibility and diplomacy in developing and maintaining cooperative external relationships
Strengthen customer relationships as an ongoing corporate imperative	Exploit fresh opportunities with a bold and unified approach		Respond rapidly to changing market conditions	
Adhere to the rules and principles of competitive strategy	Encourage personal and professional growth			

employ responsible people who are self-motivating and self-disciplined. In turn, they are rewarded with freedom, so that little energy is spent on enforcing strict procedures, hours of work, or days of vacation. Instead, emphasis is placed on what gets done. At Netflix, freedom is further expressed by building a sense of responsibility among employees, whereby they are committed to grow the enterprise.

Therefore, aligning your business strategies with your corporate culture is precisely what will give your plan the singular quality of uniqueness. In practical day-to-day terms, synchronizing strategies with the corporate culture will directly impact the markets you focus on, the image you project in the marketplace, the products and services you deliver, and the competitors you are able to face up to successfully.

From a leadership and managerial viewpoint, if you know how to identify the power of your corporate culture, you will get an unmistakable signal whether your strategies can work under adverse competitive conditions. It also provides useful insights about the success or failure of your entire business plan.

The above points are summarized as follows:

- If you are a senior executive who consciously integrates business strategy with the organization's culture, you are more likely to succeed.
- If you are a middle manager at the division, department, or product-line level who knows how to write a business plan that builds on the subculture of your business unit, you are more likely to win.
- If you recognize that corporate culture envelops your entire organization, including the caliber of leadership, the vision that drives the business plan, the boldness or timidity of strategies, the commitment to customers' needs and problems, and the care and treatment of employees, then you are more likely to succeed.
- If you do not internalize how the culture of the organization interweaves with today's hotly contested markets and do not know how to align your strategies accordingly, the results can prove fatal.

The general who understands war is the minister of the people's fate and arbiter of the nation's destiny.

Therefore, the enlightened ruler is prudent and … is warned against rash action. Thus, the state is kept secure and the army preserved.

Sun Tzu

QUALITIES OF HIGH-PERFORMING BUSINESS CULTURES

Corporate culture is part of the DNA that permeates the day-to-day organizational life of your employees. As such, it functions as the critical lifeline to your organization's future—or, as expressed by Sun Tzu, the "minister of the people's fate" that assures the "state is kept secure."

Corporate culture also delineates the types and range of strategies you can realistically undertake, which when taken to their extremes result in growth or retreat, viability or stagnation, or in its outermost limits, survival or bankruptcy.

To fully grasp the underlying nature of your organization's culture and to internalize what makes your organization tick—as illustrated in the example above of the Netflix culture—is to foretell whether your plans have a reasonable chance of succeeding. Accordingly, take the time to sort through the core values, deep-seated beliefs, and historical traditions that shape your organization's culture.

Such awareness is the primary step in formulating a business strategy. Doing so also strengthens your ability to engage the minds and hearts of the personnel who must take responsibility for its implementation. Consequently, as a tangible outcome of that effort, you will be able to develop more exacting business strategies and tactics that can win in hotly contested markets.

The following case illustrates these points.

General Electric Company CEO Jeffrey Immelt declared, "I'm intense about our competition. But I'm more concerned about our culture and our people." He further defined his concern when he admitted to two fears: first, that GE would become boring; second, that his top people might act out of fear, meaning that some executives would shy away from taking the essential risks needed to propel the company forward.

From the time Immelt succeeded the legendary Jack Welsh as CEO, he pushed for a cultural revival by driving his people to focus on creativity, imaginative marketing, innovation, and risk taking. That did not mean totally revamping GE's culture. After all, the company became a worldwide leader by adhering to Six Sigma, continual improvement of operations, cost cutting, and deal making. However, Immelt knew that in a slow-growing domestic economy and a volatile global marketplace, he had few alternatives other than to go on the offensive and push boldly into new products, services, and markets. Taking a less risky approach would mean a backward slide from which it would be difficult to recover.

How, then, is a corporate culture overhauled, where its beliefs and prac-tices have been the hallmark of excellence and the company remains the envy of executives from other high-profile companies?

The following three steps summarize GE's cultural shift.

First, compensation is linked to new ideas and customer satisfaction, with less emphasis on bottom-line results. It is based on managers' abili-ties to improve customer service, generate cash growth, and boost sales, instead of simply meeting profitability targets.

"Immelt has launched us on a journey to become one of the best sales and marketing companies in the world," says one senior GE executive. Top executives hold phone meetings every month and meet each quarter to discuss growth strategies, think up ways to reach customers, and evaluate new ideas.

Second, executives must go after businesses that extend the boundaries of GE. More than just paying lip service to the order, they must submit at least three *Imagination Breakthrough* proposals per year for evaluation and possible funding. The criteria for submitting the proposals must include taking GE into new lines of business, geographic areas, or customer groups. For some executives this approach is somewhat unsettling and goes counter to GE's former culture, which was built on nurturing internal efficiencies and generating favorable financial numbers. Notwithstanding the difficulty in shifting the mind-set toward embracing risk, Immelt sup-ports the change by fine-tuning internal operating systems to make it happen. In other words, the risks are shared and are now culturally indig-enous to the firm.

Third, executives are rotated less often, and more outsiders are brought in as industry experts instead of professional managers. That is a big departure from GE's promote-from-within tradition. Immelt pushes hard for a more global workforce that reflects the markets in which GE oper-ates. He also encourages GE's homegrown managers to become experts in their industries rather than just experts in managing.

> It is even more ridiculous when we consider that these very critics usually exclude all moral qualities from strategic theory, and only examine mate-rial factors. They reduce everything to a few mathematical formulas of equilibrium and superiority, of time and space limited by a few angles and lines. If that were really all, it would hardly provide a scientific problem for a schoolboy.
>
> **Clausewitz**

Although several of the changes at GE seem to deal with current issues related to the economy and "material factors," in fact the moves aim at revamping the culture and "moral qualities" by recasting the company for years to come. That means holding on to the fundamental beliefs, traditions, and values that embody the soul of the company and which still remain intrinsic to growth.

Similarly—depending on your level of authority—you can make minor or major adjustments to your firm's or business unit's culture. The acid test of any changes you undertake, or recommend, is that they form a logical and workable fit with the objectives and strategies of your business plan.

Accordingly, the overall culture of an organization and its individual business units embodies the core beliefs and values that drive business decisions, generate customer loyalty, and ignite employee involvement.

Be mindful, however, when you grant authority to others to make business-related decisions that they internalize the intrinsic values that drive corporate culture and consider their impact on business strategy.

> Those skilled in war cultivate the Tao and preserve the laws and are therefore able to formulate victorious policies.
>
> The Tao is the way of humanity and justice. Those who excel in war first cultivate their own humanity and justice and maintain their laws and institutions. By these means they make their governments (organizations) invincible.
>
> **Sun Tzu**

Corporate culture, as the backbone of competitive strategy, builds on the following six propositions that will directly impact your performance—and your ability to "formulate victorious policies."

1. The unique DNA lodged in every organization has a direct bearing on how you shape strategies to maintain a competitive presence in the marketplace.
2. When challenged by rivals bent on grabbing your market share or threatening your overall position, the effectiveness of your counteractions relies on your organization's prevailing culture.
3. The range of available options you can use to differentiate your product, create value, and satisfy customers is governed by the totality of your organization's culture.

4. Correctly interpreting the deep-rooted values, behaviors, and traditions that drive your organization and inspire your employees can significantly affect the success or failure of your business plan.
5. Even if you are not in a position to alter your organization's culture or business unit's subculture, you should know how to select competitive strategies that will likely gain support within the prevailing value systems that drive the organization.
6. Sensitivity to your organization's beliefs and patterns of behavior acts as both an indicator of past performance and a predictor of what the organization—and you—will achieve in the future.

He selects his men and they exploit the situation. Now, the valiant can fight; the cautious defend, and the wise counsel. Thus there is none whose talent is wasted.
Do not demand accomplishment of those who have no talent.

Sun Tzu

What type of corporate culture will it take to "exploit the situation"? Here are some insightful comments from heads of leading organizations that may shed some light:

A business can become stronger by making itself a community of people who share the same ideals and goals, the same corporate culture.

CEO, Disco Corporation

Sharp has a heritage of creating one-of-a-kind products. It is part of our corporate DNA.

President, Sharp Corporation

Innovative ideas are born of bold dreams and beliefs, and energized through inspired technology and a clear vision.

President, Matsushita Electric

My personal credo of three Cs: Challenge, Create, Commit. I tell all my staff to approach life with a pioneer spirit—several steps ahead of the competition.

President, Itochu Corporation

We have many challenges ahead of us, but perhaps our biggest challenge is the one we have created for ourselves. I mean growing Toyota into a company that truly matters to our customers, our employees, and to the societies where we live.

CEO, Toyota Motor Corporation

Now the method of employing men is to use the avaricious and the stupid, the wise and the brave, and to give responsibility to each in situations that suit him. Select them and give them responsibilities commensurate with their abilities. Do not charge people to do what they cannot do.

Sun Tzu

You can achieve excellence as a leader when your people are disciplined and committed to the organization's cultural values. Excellence in leadership, however, does not mean perfection. On the contrary, an excellent leader allows subordinates room to learn from their mistakes as well as from their successes—as long as "you give them responsibilities commensurate with their abilities."

In such a positive climate, people work to improve and take the risks necessary to learn. In contrast, a manager who sets a standard of zero defects, no mistakes, is also saying: *Don't take any chances. Don't try anything you can't already do perfectly, and don't try anything new.*

For instance, *Twitter Incorporated* co-founders Biz Stone and Evan Williams created a positive climate as they looked to grow beyond the phenomenal introduction of their micro-blogging service. As part of their strategy, they emphasized growing Twitter slowly in order to find people who fit with its culture. The culture for Stone and Williams is partially defined through a variety of rituals, such as family-style lunches and "happy hours."

Expressed differently, Stone and Williams had to reach the hearts as well as the minds of employees. It is *heart* that collectively describes the underlying emotional qualities by which they would lead. Actively reaching out to win employees' hearts meant encouraging them, conveying confidence in their work and attitudes, offering appreciation, and wherever possible, providing tangible security through compensation as well as through meaningful training.

Heart, then, is the cultural pathway that separates stability from uncertainty, enthusiasm from discouragement, and courage from fear. These behavioral qualities govern the hearts and minds of individuals. Alternatively, where positive values, beliefs, and traditions are ignored, abused, misinterpreted, or carelessly altered, the organization's culture is violated, and *down* comes the instigating manager, which will negatively impact the firm's product line and market position—and even the stability of the entire company. *Up* goes the alert competitor who sees the disruption and enters the confusion with well-timed actions to fill a market void—provided its culture is tuned in and responsive to such competitive opportunities.

In part, the above characteristics are explained in the overconfident attitude of *we've got the market locked up*, which often translates to complacency and ends with defeat. Such feelings tend to dry up creativity and stifle the inner drive to push for excellence. It draws off the adrenalin that comes from fighting within a performance-based marketplace.

The marketplace is littered with one-time leaders in such fields as automobiles, electronics, metals, and consumer goods. Some companies in those industries turned around at the last moment. Others were acquired or went bankrupt.

The mind-set was usually supported by the inaccurate notion that high market share means security, which was often followed by a passive defense of that market position and the attitude of being too entrenched in the market to fail. Those conditions at one time or another faced such high-profile companies as *Xerox, EMC,* and *Hewlett-Packard*. All, however, have aggressively pulled out of the slump—or are in the process of doing so.

> If an army has been deprived of its morale, its general will also lose his heart. Heart is that by which the general masters. Now order and confusion, bravery and cowardice, are qualities dominated by the heart.
>
> Therefore the expert at controlling his enemy frustrates him and then moves against him. He aggravates him to confuse him and harasses him to make him fearful. Thus he robs his enemy of his heart and of his ability to plan.
>
> **Sun Tzu**

How a business strategy evolves and is ultimately implemented against a competitor generally has the imprint of the individual manager. Yet to succeed, the "ability to plan" must incorporate the innate cultural characteristics of your firm. In turn, it has to embody the "qualities dominated by the heart" of those individuals who must live day-in and day-out with their decisions and ultimately with the good and bad consequences.

And where these same qualities affect your performance, they also concern your competitor's effectiveness, so that your aim is to be aware of what you can do that "rob his ... heart and his ability to plan." Competitive strategy and human behavior, therefore, are fused tightly to corporate culture through two tangible applications. First, it forces you to take a broader strategic view of how to approach markets. It also guides your tactical decisions when launching a new product or service, entering a new market, or expanding an existing market. Second, it determines how bold or passive your business plans are likely to be against competitors. See Table 5.2.

> Now when troops flee, are insubordinate, distressed, collapse in disorder, or are routed, it is the fault of the general. None of these disasters can be attributed to natural causes.
>
> When troops are strong and officers weak the army is insubordinate.
>
> When the officers are valiant and the troops ineffective the army is in distress.
>
> When senior officers are angry and insubordinate, and rush into battle with no understanding of the feasibility of engaging and without awaiting orders, the army is in a state of collapse.
>
> When the general is morally weak and his discipline not strict, when his instructions and guidance are not enlightened, when there are no consistent rules to guide the officers and men and when the formations are slovenly the army is in disorder.
>
> When any of these conditions prevail the army is on the road to defeat. It is the highest responsibility of the general that he examines them carefully.
>
> **Sun Tzu**

As you begin aligning your strategies with your corporate culture, one remaining issue requires your attention: *beliefs*. Beliefs are convictions

TABLE 5.2

Guidelines to Align Business Strategies with Corporate Culture

Would your corporate culture permit you to do the following:

Implement strategies bold enough to frustrate your competitor's plans, or would they be too limiting to do the job?

Disrupt your rival's alliances and thereby weaken the impact of the opposing manager's strategies?

Unbalance the competing manager into making tactical mistakes?

Spread the seeds of uncertainty and doubt among the competitor's personnel through aggressive actions and disinformation that "rob his ... heart and his ability to plan," even before the tough marketing battles begin?

employees hold as true. They are anchored to the attitudes, values, and mind-sets that usually originate at the senior executive level and filter down through the organizational layers to shape their employees' behavior and performance.

Sun Tzu broadens his viewpoint of behavior and performance to include such negatives as undisciplined behavior, morally weak leadership, insubordination, disorder, and lack of discipline that can cause "a state of collapse." He maintains these issues should be the "highest responsibility of the general that he examines them carefully." Sun Tzu further declares, "When any of these conditions prevail the army is on the road to defeat."

Today, as many traditional organizational hierarchies flatten and responsibilities are delegated to the lower echelons, executives in most well-run organizations take very seriously the notion that employees are free to feed off their personal beliefs. They also take into account the widespread and persuasive influence of messaging through social media, which can alter even the most ardent beliefs. All of which puts managers on notice to stay prudently alert to employees' overt and hidden behaviors.

Ultimately, if not monitored, undetected changes can affect your ability to successfully implement your plans. Beliefs, therefore, can unceremoniously and unconsciously creep into employees' thinking and impact their behavior—and ultimately influence your ability to lead.

Diversity is still another category that impacts beliefs. Beliefs reflect individuals' upbringing, heritage, religious practice, and cultural traditions. In turn, beliefs influence in such distinct ways as apathy or support of a new service, drabness or originality in designing a unique product, and courage or fear in taking on an aggressive competitor.

A company's weakness or strength comes from such diversity. Your obligation, then, is to recognize, respect, and harness different backgrounds and personal convictions, as long as they do not conflict with the core values of your organization.

Your task as a leader is not only to acknowledge that people are different, but also to value them because of their differences, given all the potential ingenuity and complexity that reside in their minds.

Further, you can take advantage of these unique differences and mobilize them into a cross-functional team. The pragmatic issue here is that diversity should be viewed as a desirable situation and that people of different backgrounds bring different talents to the table (see Table 5.3).

Cargill, the agricultural and food-ingredient giant with more than 100,000 employees in 59 countries implemented its first diversity

TABLE 5.3

The Functions and Responsibilities of a Cross-Functional Team

Functions include

Define the business or product-line strategic direction, also known as vision and mission.

Analyze the environmental, industry, customer, and competitor situations.

Develop short- and long-term objectives and strategies.

Prepare product, market, supply-chain, and quality plans to implement competitive strategies.

Responsibilities include

Create and recommend new or additional products and services.

Approve all alterations or modifications of a major nature.

Act as a formal communications channel from the market back to internal departments.

Plan and implement strategies throughout the product life cycle that utilize competitor intelligence to determine indirect approaches that can be put into action with speed at a decisive point.

Develop programs to improve market position and profitability.

Identify product and service opportunities in light of changing consumer buying patterns.

Coordinate efforts with various corporate functions to achieve short- and long-term objectives.

Organize efforts for the interdivisional exchanges of new market or product opportunities.

Develop a strategic business plan aligned with the corporate culture.

---

blueprint in 1995. The program, known as *Valuing Differences*, focuses on integrating the principles of diversity and inclusion into the company's performance-management system.

"Our belief is that diversity is not just about getting people in the door, you must have the right culture and an environment that allows everyone to contribute their fullest to reach their potential. Equal access for promotion and opportunities is a major part of our diversity effort. It is much more than just recruiting," states Cargill's vice president of corporate diversity.

The highlights of the Cargill program include the following:

- An annual survey to measure response to its diversity initiative, giving the company a baseline index of how well the initiative works.
- Each business unit writes its annual contract with Cargill's chief executive to outline its tasks and objectives. The contract includes *Valuing Differences* as performance criteria.

- A mentoring system in which junior-level lesbian, gay, bisexual and transgender (known collectively as LGBT) employees volunteer to mentor more-senior managers for a year.

Carried to its extreme, diversity is further magnified by the new technologies that can marshal the talents and inputs of millions of people worldwide. For instance, search engine *Google* instantly polls millions of people and businesses whose websites link to each other, and companies such as *Procter & Gamble* and the *LEGO Group* use Internet-powered services to tap into the collective beliefs of employees, customers, and outsiders, which are then used to transform their internal operations and product development activities. Procter & Gamble now gets 35 percent of new products from outside the company, up from 20 percent three years ago. That has helped boost sales from research and development by 40 percent. LEGO uses the Internet to identify its most enthusiastic customers to help it design and market more effectively. After a new locomotive kit was shown to just 250 train fans, their word-of-mouth through the Internet helped the first 100,000 units sell out in 10 days with no other marketing.

Consequently, if you take into account the expansive role beliefs play in preparing employees to develop and implement a business strategy, you will enhance your chances of achieving objectives. Employees often act and win over tremendous odds when they are convinced of the ideals and beliefs for which they are working.

Therefore, be aware of beliefs and culture in three contexts: First, look to enhance your sensitivity to the diverse backgrounds of your employees, as well as to relationships within the supply-chain. Second, increase your awareness of the active culture that is exhibited in your markets. Third, secure a lifeline to your customers by understanding their customs and traditions, and then incorporate the significant ones into new product features at the beginning stages of development.

The examples of *Cisco Systems* and *Charles Schwab & Company* illustrate the irrefutable relationship of business strategy and corporate culture. Cisco Systems is characterized by a near-religious convergence on the customer, a total belief in employees as intellectual capital, and an energetic willingness to team up with outsiders to develop active partnerships. This passion for molding such an outside-in focus is credited to the leadership of CEO John T. Chambers, who clearly saw those attributes as a value system to drive all subsequent actions. In turn, other levels of employees recognized such values and used them to shape strategies that conformed to the culture of the organization, thereby increasing the chances of successfully implementing business plans. Further, by accepting those characteristics as a functional formula, line managers were able to reliably predict the success or failure of their respective strategies—depending, of course, on the competitor's ability to challenge them, and provided they correctly interpreted the competitor's culture.

In response to a reorganization at Charles Schwab & Company, employees initially resisted taking brokerage orders over the Internet. Senior management's approach to change was to assemble nearly 100 managers at the base of San Francisco's Golden Gate Bridge and hand each a jacket imprinted with the slogan, *Crossing the Chasm*. The group proceeded to march across the bridge and symbolically walk into the Internet Age. That physical act symbolized the cultural reinvention of the company and the strategy of burning bridges behind them.

Value systems and practices are often wrapped in symbols and rituals as ways to initiate and sustain change. The following is what you should know about symbols and rituals to enhance your managerial effectiveness.

> I make the enemy see my strengths as weaknesses and my weaknesses as strengths, while I cause his strengths to become weaknesses and discover where he is not strong.
>
> I conceal my tracks so that none can discern them; I keep silence so that none can hear me.
>
> **Sun Tzu**

Symbols and Rituals

There are strategy implications to symbols. Used as signs, acts, or objects, they can signify special meanings, as well as convey messages that "make the enemy see my strengths as weaknesses and my weaknesses as strengths." In turn, they shape the plans within your organization, just as they communicate signals to those rivals observing from the outside.

Therefore, if an organization is to remain a unique entity, it can be represented symbolically through a dress code, an oath, a song, or the manner of addressing people with certain terms or titles.

It carries further to the choice of words that represent boldness or weakness of a business strategy, as well as to the expected behavior when entering or defending a market. What follows, therefore, is that an employee's allegiance to the organization is expressed through symbolism.

Connected to symbols are rituals. These consist of traditional or contrived ceremonies (as in Schwab's *crossing the chasm*) in which some physical act or expressive behavior dominates over otherwise technical or rational actions.

People tend to think of organizations as physical units and part of the material world. Yet, the reality is that rituals represent the means by which people are linked to organizations to give it a *human* dimension and thereby a differentiated quality.

Because rituals assume various forms, it would be highly useful to investigate the meanings, types, and structures of the symbols used in your company's rituals. Then stay alert to your employees' beliefs in the effectiveness of the rituals. Where accepted and meaningful, it gives employees confidence, dispels their anxieties, and disciplines their work groups.

In some situations, rituals may do nothing more than tighten the relationships between one business unit and another. Or they may be used to bring together diverse units toward a unified corporate goal. Again, in the Schwab case it was expressed by *crossing the chasm*.

You will also find cultural symbols and rituals most useful when viewed as a form of communication through the spoken word. Words define and interpret what is really going on in the company. Hence, they embody each person's behavior through the social networks that exist among the various individuals who are interacting. The following examples illustrate the various points associated with symbols and rituals and their resulting impact on initiating cultural change within the organization.

GE Aircraft Engineers Division at one point shifted its engineers into a renovated warehouse that had the look and feel of a high-energy start-up. The aim was to redefine an existing culture from a slow-moving, tradition-based establishment to one that emphasized creativity and ingenuity. The symbolism associated with the physical layout formed the basis for change.

Heineken Brewery initiated a cultural change at the tradition-bound 400-year-old beer maker. The aim was to stir employees out of their complacency and push Heineken to break away from its play-it-safe corporate culture. Traditional corporate symbols and rituals were shaken by the bold strategy of making a dozen acquisitions, moving aggressively into seven countries in eastern Europe, capturing the sought-after twenty-something segment, and introducing daring new packaging.

The following lessons surface from the above examples: First, it is in your best interest to pay close attention to your firm's value systems, your employees' deep-rooted beliefs, traditions, and symbols when determining your organization's future direction. Second, if you have bottom-line responsibilities, your everyday job will be more productive if you match your business plans for a market segment, product line, or sales territory with the unique culture of your business unit. Third, employees responsible for implementing strategies must be oriented toward culturally diverse markets, remain flexible, and be adaptable to change. They must tune in to the nuances of the markets and be receptive to their unique value systems, beliefs, and forms of behavior.

> He whose ranks are united in purpose will be victorious. The appropriate season is not as important as ... harmonious human relations.
>
> **Sun Tzu**

REVITALIZE YOUR COMPANY'S CULTURE

Revitalizing your company's culture requires the same degree of attention as you would give to any major initiative. First and foremost it means developing in your staff an attitude of "united in purpose" (as described above for Netflix) and fostering "harmonious human relationships."

Disregarding that suggestion, especially in today's hotly contested markets, can prove problematic. That is, even if you devise a brilliant strategy, attempting to implement change with any measure of success without initially sorting through the core values, beliefs, and traditions that mold your organization can affect the outcome.

Once you have taken steps to unify your group in support of the business plan, you can use the following guidelines as part of the process to revitalize the corporate culture and align it with your strategies:

- Stay on the offensive.
- Facilitate boldness, depending, of course, on the availability of resources and the level of confidence you show in employees, as well as the confidence they show in you. If the risk succeeds, offer ample rewards; if failure results, avoid damaging repercussions.
- Encourage creativity and innovation.

- Seek maximum input from employees at all levels of the organization. Try new ideas that could lead to new products, evolving markets, or new business. Maintain a cultural sensibility that retains an open mind and avoids the idea-killing attitude that *we've tried that.*

- Allow sufficient time for ideas to incubate and hatch into new technologies, products, and services. Develop concrete formats for employees to submit ideas. The case of *General Electric*, described above, that required executives to submit three *Breakthrough Imagination* proposals each year is a good example.

- Learn to live in a flexible competitive environment. This is a cultural attribute that is often difficult to embed within an organization, and equally difficult to instill in employees. It also needs senior management's full support, especially during this period of severe market volatility.

- Maintain an outward display of resolute calmness and unshakable confidence. For some employees, extreme change creates an unsettling situation where any perceived upheaval in conditions in or out of the organization is difficult to endure. Still, flexibility is a singular characteristic that must be maintained. As such, it is an imperative for operating successfully in the Internet Age.

- Act as an aggressive competitor. This combative mind-set helps you discover where your firm has a comparative advantage and where it is at risk. It indicates strengths and weaknesses in your products, services, logistics, and overall organizational structure. You also gain insight by examining relationships with suppliers, intermediaries, and customers along the entire supply-chain. The process exposes strong points and vulnerable areas in technology, manufacturing, human resources, and capital resources, and it surveys any other area that might endanger your firm to competitive attacks or prevent you from taking advantage of ripe opportunities. As important, it unmasks sensitive information on employee behavior and suggests clues on how to undertake change. Such exposure also sheds light on those senior executives who cannot (or choose not to) make determined efforts to take on an aggressive posture. In practical terms, few executives are effective for all seasons. And not all individuals are capable of performing optimally through successive stages of a corporate cycle—start-up, growth, maturity, and decline—or even within different cultural environments.

The natural goal of all campaign plans therefore is the turning point at which the attack becomes defense. If one were to go beyond that point, it would not only be a useless effort which could not add to success. It would in fact be a damaging one.

What matters, therefore, is to detect the culminating point with discriminative judgment.

Clausewitz

- Build a solid market position. The aim is to create a unique market position from which competitors cannot easily dislodge you. To the extent you are able, try to create brand equity and brand recognition. Your managerial efforts should be directed toward mounting a long-term positive image for your firm. Research has indicated that high market share equals high return on investment. Some executives go so far as to advise continuing to maximize market share at any cost. This viewpoint, however, remains controversial among some executives and academic scholars. Clausewitz's point of view is, "going beyond the turning point ... could not add to success" and "what matters is to detect the culminating point with discriminative judgment." Others believe that chasing market share no longer guarantees profitability. One point that is not controversial is that customer satisfaction and long-term customer relationships remain the enduring principles.
- Stay close to evolving technology. Tune in to what is happening in those technologies that can help transform your business to the new economy models. (Look again at the Cisco case.) Several choices exist: buying a technology, investing in start-ups, or partnering with a compatible company.

You can employ all or some of these steps to drive cultural change and energize your company. How drastic those changes are depends on the severity of your company's problems. Therefore, you can react to problems as they arise or you can be proactive by anticipating changes. It is all part of leadership and managerial competence, which relies heavily on estimating your internal and external environments, including competitors, suppliers and, most of all, your customers.

In addition to the above criteria, Table 5.4 further defines the components and resulting benefits that energize a healthy corporate culture.

TABLE 5.4

Energizing a Healthy Corporate Culture

Attributes	Benefits
Diversity	A company's strength comes from its diversity, where respect prevails for different backgrounds—as long as they do not conflict with the core values of the organization.
Fair treatment of employees	Employees support company efforts as long as equality exists—and where rewards and disapprovals are applied consistently.
Pride and enthusiasm	Employee zeal spills over to business partners and customers.
Equal opportunity for employees	A heightened spirit of innovation helps employees achieve their full potential, leads to team cohesiveness, elevates morale, and encourages innovation.
Open communication	Provides a channel to pass on beliefs, values, and unite personnel toward a vision for the future of the organization.
Respect for employee contributions	Enhances involvement and enthusiasm to work toward common goals and strategies.

In war the result is never final.

Never assume that … its whole existence, hangs on the outcome of a single battle, no matter how decisive. Even after a defeat, there is always the possibility that a turn of fortune can be brought about by developing new sources of internal strength.

Clausewitz

SUMMARY

The consensus among many executives is that the purpose of corporate culture is to develop an internal work environment that encourages individuals to perform efficiently. Embedded in that belief should be a total awareness that "even after a defeat, there is always the possibility that a turn of fortune can be brought about by developing new sources of internal strength."

What follows, therefore, is to revitalize your corporate culture so that it is aligned with the organization's vision, objectives, and strategies, all of which are powered by "harmonious human relations," and tuned to the dynamic forces that make up the competitive marketplace.

Cisco Systems, cited earlier, embodies such a culture with its intense attachment to customers, a total belief in employees as intellectual capital, and an energetic willingness to team up with outsiders to develop active partnerships.

By turning to corporate culture and using it as an additional managerial tool, you derive a significant competitive advantage. From a people viewpoint, you achieve high employee motivation and increase team cohesiveness. You, thereby, create an ironclad connection between competitive strategy and corporate culture.

6

The Force Multiplier behind Your Business Strategy: Leadership

A commander need not be a learned historian or a pundit, but he must be familiar with the higher affairs of state and its most intimate policies. He must not be an acute observer of mankind or a subtle analyst of human character, but he must know the character, the habits of thought and action, and the special virtues and defects of the men whom he is to command.

This type of knowledge cannot be forcibly produced by an apparatus of scientific formulas and mechanics. It can only be gained through a talent for judgment and by the application of accurate judgment to the observation of man and matter.

Clausewitz

Clausewitz's all-encompassing statement casts a leader as knowledgeable of the strategic direction of the business, one who is clued in to the operating culture that drives the organization, a leader who is intimately aware of the staff's moods, behaviors, and actions. He sums up by stating that a leader requires "the application of accurate judgment to the observation of man and matter."

To give a manageable framework to such wide-ranging traits of leadership, the following applications provide pragmatic guidelines to Clausewitz's enduring points: *strategic direction and policies, self-confidence and leadership, mastering leadership skills, barriers to effective leadership, leadership in the competitive world,* and *using leadership as a force multiplier.*

STRATEGIC DIRECTION AND POLICIES

Strategic direction is the juncture at which you "must be familiar with the higher affairs of state." It is where your experience, skill, and insight converge to envision the future of your organization or business unit. It is the point where you take a broader view of the internal and external issues affecting your business. That way you can provide greater precision to prioritizing objectives, shaping competitive strategies, and deploying people, material, and financial resources for maximum impact.

An indispensable dimension in thinking about the overall direction and goals of the organization is that it binds your leadership into a cohesive and managerial whole. That is, you and your staff can come together and make sense about the workings of the organization and the dynamics of the marketplace. You thereby create order out of what can easily deteriorate into disorder due to the lack of information and understanding about the course of the company.

What follows, as you assign authority and responsibility to members of your staff, is a noticeable improvement in the skillful implementation of a business plan and a heightened level of morale. In turn, that goes far to give a human face to your leadership. Altogether, you are able to energize the rank and file to push forward in an increasingly competitive environment. Often, it is the singular factor in deciding the success or failure of the business plan.

> One who is confused in purpose cannot respond to his enemy.
>
> **Sun Tzu**

This coming together to shape a strategic direction is known as collaborative, community-based, or social strategy planning.* The process is beginning to take hold among a growing number of organizations, including *3M*, Dutch insurer *AEGON*, global IT services provider *HCL Technologies*, Linux software provider *Red Hat*, and defense contractor *Rite-Solutions*.†

As you begin to take on a strategic viewpoint, you tend to activate your mind and think of resourceful approaches to improve the financial health

---

* Appendix 3 provides a format for a team-based strategy action plan.
† McKinsey & Company research, collected over more than a decade at some 600 companies, indicates that even at the healthiest companies, about 25 percent of employees are unclear about their company's direction. That figure rises to nearly 60 percent for companies with poor organizational-health scores.

and competitive vitality of your organization. The following points indicate how a broad strategic awareness can impact day-to-day operations.

You are able to stay informed and maintain better control especially where individuals are geographically separated and directives from higher-ups are often misinterpreted. You can then gain a clearer perspective about changes taking place in diverse markets, such as trends in consumer behavior, competition, environment, and culture. Armed with this information, you are in a better position to draw meaningful conclusions about the possible impact of various market forces on your company. And you gain broader insights into potential markets and businesses your company should be looking into over the next several years.

As you learn more about the additional functions required to fulfill customers' needs as the market evolves, you acquire a better sense of your organization's current competencies and what new skills and functions are needed to sustain growth. You can then tailor the appropriate skills training at the critical junctures to exploit market opportunities as they arise.

Further, you are better able to deploy personnel according to market conditions and according to their experiences and personality traits, so that when you create teams of personnel. The more aggressive individuals will act boldly but not impulsively; and the more reticent ones will act with some bravado, yet not retreat prematurely under competitive pressure.

As you expand your knowledge about the categories of customers you will serve, you can contribute valuable insights to your organization in such decision areas as new technology investments, joint-venture options, product development recommendations, supply-chain issues, and more.

The essential point is that awareness of an organization's strategic direction no longer belongs exclusively to senior management; it is essential for effective and responsible leadership at all operating levels through collaboration.

Discipline does more in war than enthusiasm.

Machiavelli

SELF-CONFIDENCE AND LEADERSHIP

A self-confident mind is not just capable of strong mental exertions. It is one that in the midst of tackling severe problems can maintain its equilibrium regardless of internal turmoil.

To expect that you will remain calm and reserved in every competitive situation is asking for a superhuman effort. The agonizing feeling of failure is not a fabrication or an illusion. It is a conscious realization, for instance, that a competitor may be superior for reasons that were unforeseen but have become disturbingly clear to you as market conditions unfold. Such conditions could legitimately cause a sudden collapse of all hope, a breaking down of self-confidence. Instead of working energetically to stem the tide, subordinates fear their efforts will be useless; they hesitate to move, and soon leave everything to fate.

The essential point is that competitive encounters assume that human weaknesses do exist. These materialize when momentary negative impressions of events distract you from your objectives. Therefore, make every effort to put such disruptions into correct perspective immediately, however uncomfortable they may appear. Keep in mind, too, as part of such a perspective that "discipline does more in competitive encounters than enthusiasm."

Also, when tempted to doubt the correctness of your decisions, keep your trust in Sun Tzu's time-tested leadership qualities and strategy principles that follow:

> If wise, a commander is able to recognize changing circumstances and to act expediently.
>
> If sincere, his men will have no doubt of the certainty of rewards and punishments.
>
> If humane, he loves mankind, sympathizes with others, and appreciates their industry and toil.
>
> If courageous, he gains victory by seizing opportunity without hesitation.
>
> If strict, his troops are disciplined because they are in awe of him and are afraid of punishment.
>
> **Sun Tzu**

MASTERING LEADERSHIP SKILLS

There is enough evidence to suggest that able leadership and outstanding managerial performance include natural and learned skills—as well as a keen inner sense "to recognize changing circumstances and to act expediently."

And should the acquired skills merge with an individual's innate abilities, the outstanding leader is clearly distinguished from the mediocre manager.

Thus, we come down to the instinctive and acquired traits of leadership. In addition to accumulated historical sources, numerous academics, consultants, military leaders, and other practitioners lay claim to understanding and teaching the qualities of leadership. What follows is a summary of the generally accepted ones. If adapted, they can enhance your skills to manage people, resources, and implement effective business-building opportunities. These include *insightfulness, straightforwardness, compassion, strictness,* and *boldness.*

> *Insightfulness.* You show the aptitude to recognize early on the impact of new competitors coming into your market space. As vital, you display a special acuity to spot changes in your customers' buying behavior, shifts in your industry's use of new technologies, and environmental trends, such as the "green" movement, on your business.
>
> *Straightforwardness.* Your employees have no doubt how and when rewards or reprimands will be handed out.
>
> *Compassion.* You respect your employees, appreciate their hard work, and empathize with them during stressful times.
>
> *Strictness.* You are dedicated to the long-term objectives of your company. In turn, your people respect your discipline and optimistic outlook.
>
> *Boldness.* You find innovative ways to market your company or product by seizing opportunities; and you show the confidence to avoid getting bogged down in morale-breaking indecision.

The commander must trust his judgment and stand like a rock on which the waves break in vain. It is not an easy thing to do.

Clausewitz

Now, we can apply those positive leadership qualities to the realities of creating opportunities and challenging competitors: First, within a dynamic and changing marketplace, you have to work with the best information available. Second, you "must trust [your] judgment" by acting boldly and implementing strategies with speed—even where data about your competitor are sketchy, and despite other uncertainties. Next, should you feel weighed down with apprehension, rely on inner support by drawing on your intuitive ability to boost your courage and by following your instincts, however faint, to find the proper path.

Some managers realize they must be strong minded; at the same time they sense the dangers of a flawed decision. They are not sure what is before them. And even if they are in the habit of acting with speed, the more they linger with the dangers of indecision, the more doubtful they become.

One reliable approach to reduce the level of anxiety is to make estimates beforehand. Therefore, use every bit of available market intelligence you can get hold of. Then develop alternative courses of action. Doing so prepares you to take clear-thinking action at any given moment; you thereby alleviate the unsettling feeling that comes with indecisiveness.

As for boldness, this quality develops into a personality attribute only if it progresses into a way of thinking. The evolving pattern means driving with single-mindedness to reaching a specific objective. There are many talented individuals who under somewhat normal conditions can logically form an appropriate solution to a problem. Yet, under sudden competitive conditions, they simply cannot come forward with the determination and courage to make tough decisions. Often, this lack of nerve is triggered by their natural anxiety associated with disappointment.

Overcoming adverse feelings relies on training, discipline, and experience. Keep in mind, too, that you are in a match of mind against mind: your mind opposing the mind of a competing manager who may be confronting similar reactions. You want to be the one who wins out and moves forward. (See Table 6.1.)

TABLE 6.1

Guidelines for Successful Leadership

Hold fast to the definitive object of all business, which according to the late management scholar, Peter Drucker, is to "create a customer."
Remove obstacles that prevent people from gaining pride in their contributions to quality service.
Reduce the negative effect of any turf-building barriers within the company by establishing collaborative cross-functional teams.
Introduce a work environment where the emphasis is on junior managers leading, not merely supervising.
Communicate long-term goals to the staff, which they can translate to such areas as developing leading-edge products, improving product quality, and introducing value-added services.
Eliminate the use of fear as a motivator.
Encourage employees to express ideas; listen to them and respond.
Promote self-improvement as an ongoing priority for career advancement.
Institute ongoing employee training and education to advance their skills—with emphasis on gaining competence in competitive strategies.

Now, the supreme requirements of generalship are a clear perception, the harmony of his host, a profound strategy coupled with far-reaching plans, an understanding of the seasons and an ability to examine the human factors.

Sun Tzu

BARRIERS TO EFFECTIVE LEADERSHIP*

The following list identifies a variety of barriers that could distort "a clear perception" of your goals and hamper your ability to clearly appraise the "human factors."

Employees Remain Deficient in Skills

If your staff lacks the skills and discipline to implement plans, the effect can have dire consequences on your ability to defend your existing markets. The long-term outlook can be equally ominous if you are unable to successfully expand your reach into new markets or take advantage of fresh opportunities.

The key point is that, even where you have diligently followed good practices in developing competitive strategies, your chances of success are limited without skilled individuals to put forth the disciplined effort that makes the difference between success and mediocre performance. What should follow is that with fine-tuned skills and discipline, your staff has a better chance of resisting the pressures from the inevitable counteractions of competitors. That said, it remains your responsibility to provide for ongoing training to strengthen the skills of those who are going to carry out your plans.

Employees' Negative Perception of Managers

This is a difficult barrier to overcome, especially if employees perceive their manager is unable to make accurate decisions. As bad, the manager does not understand the potential threats—and is even unaware of missed

---

* This section overlaps some of the content in Chapter 2, where effective leadership is an essential component in securing a competitive lead through speed of action.

opportunities. In turn, that realization can lead to a disturbing morale problem among the staff, which is another one of the "human factors" that can magnify further into a total loss of confidence.

What follows is that the negativity can transform into feelings of hopelessness, ranging from job loss to threats against the organization. Thus, if you observe any hint that your staff is losing confidence—or worse yet, they think the organization may be losing its way, you may have to deal with the negative consequences.

Therefore, make every effort to keep employees tuned in to evolving events, listen to their comments and gain a sense of their conversations, verify and flesh out information that is true, uncover false information and immediately squelch unfounded rumors with qualified information. Then set up a forum whereby employees offer their input and get them involved in the reasoning process that can lead to possible solutions, that is, strategies.

Limited Support from Management

This obstacle can signal poor internal communications. Where such a gap exists, there is limited control. Individuals at the lower echelons of the organization feel left out and sense they have to fend for themselves. Although there are instances where some groups may relish the idea of being left on their own, the overwhelming majority need to know that senior management is fully aware of their plans and strategies.

They need to know that timely support is available in such areas as approving investments, shifting resources to secure a favorable competitive position, launching a product, or improving customer services. In addition there are the sudden bursts of activity from a competitor that could be blunted if backup were provided rapidly. Here, again, open communication through any number of technology forums and face-to-face meetings can eliminate the barrier.

Conflicts Concerning Objectives, Priorities, and Strategies

Here is the perennial issue that can form an obstacle to effective leadership. This is especially so where individuals in your group cannot agree on prioritizing objectives and agreeing on strategies. Of course, as a senior manager you can arbitrarily override conflicts and make the choices.

If that approach is taken, you may have to mend hurt egos and placate others. Where you act as a mediator and let the individuals resolve their issues, you still have to deal with the hard decisions, especially where there are budgetary constraints and limited resources. In such instances, you will still have to intervene and decide among many business proposals, all of which are vying for attention.

Inevitably, some plans are going to get dropped or modified, as others get the go-ahead. In the end, leadership is accepted or rejected, as shown above, on the perception your staff has about your ability to make effective decisions.

> Keep one's head at times of exceptional stress. Strength of character does not consist solely in having powerful feelings, but in maintaining one's balance in spite of them. Even with the violence of emotion, judgment and principle must still function like a ship's compass.
>
> **Clausewitz**

A Ponderous Corporate Culture

This can be a daunting barrier. Yet it is one that you must face as a leader if you are to "keep [your] head at times of exceptional stress." It is discussed at length in Chapter 5. If the company's culture places a drag on getting your projects moving, or if it is functioning at too slow a pace to keep up with changing market conditions, then you have choices.

First, if you are in a position to make a wholesale change in the corporate culture to keep in tune with the prevailing conditions in your industry, then you have a decided advantage to get the organization in sync with the competitive environment. Whereas, if you are not in a senior position to make changes, your best approach is to adapt your plans to the existing culture. Second, you can attempt changes in the subculture, meaning a smaller business unit of the organization. (Review Chapter 5 for techniques to make changes.) This pathway is probably more doable for most managers.

The key point is that whichever approach you take, you should make certain there is an alignment of your business plan with the corporate culture.

Limited New Product Development

Another major obstacle to strengthening your leadership role is an inability to initiate and oversee the flow of new products. Your company's growth depends on new products. Sales people clamor for new ones. And customers wait for the winning ones.

Even if your position is not directly related to new product development, you still can, through your enhanced role as a leader, initiate proposals leading to new product development.

What is meant by a new product? One accepted market-based definition states: A product is new when it is *perceived* as new in the marketplace, and not when your company decides a product is new. If you accept that viewpoint of perception, you have several options open to you: Beyond totally new-to-the-world innovation, new products can range from modification or line extension to diversification. Each requires changing the product either slightly or extensively. Also, as you consider new products, look at the opportunities for remerchandising and market extension. These last two categories do not alter the product but permit a perception of a *new* product.

Table 6.2 presents the differences among these five categories of new products from the viewpoint of perception. Rarely will the five categories of new products presented in the table be used separately. They lend themselves to combined applications for maximum impact. For instance, line extension is often used with remerchandising or market extension. Diversification is often combined with market extension.

TABLE 6.2

Categories of New Products

Category	Definition	Nature	Benefit
Modification	Alter a product's features, add-on services, add functionality to packaging	Same number of product lines and products	Combines the new with the familiar
Line extension	Add more variety	Same number of product lines, higher number of products	Segments the market by offering more choices that are perceived as new
Diversification	Enter a new business	New product line, higher number of products	Spreads risk and capitalizes on opportunities
Remerchandising	Market change to create a new impression	Same product, same markets	Generates excitement and stimulates sales
Market extension	Enter a new market with applications	Same products, new market	Broadens the base

For increasing degrees of marketing newness, you can differentiate between remerchandising and market extension.

The use of one category does not preclude the application of other approaches at the same time, possibly within the same market. What remains essential, though, is that the prospective customer perceives a difference worthy of consideration.

Unwieldy Committees That Initiate Delays

Another common barrier to effective leadership occurs when a market condition that warrants timely action and approval is not forthcoming. If an unwieldy committee is the problem due to extensive deliberation that reaches a point of procrastination, clearly there is an obstacle. This is not to say that committees are unnecessary; the issue is where committees lose their focus, objectives are not clearly stated, urgency is misread, and little is accomplished.

What you want to avoid are those hurdles that prevent you from getting effective decisions when needed. Fluidity is what you want to achieve. Each obstacle, such as a ponderous committee, can severely alter your ability to lead with confidence and purpose.

Pressure from Aggressive Competitors

The fundamental problem here is that sudden and aggressive actions by competitors can strike fear among your employees, resulting in damaged morale and lost momentum.

Such actions can leave deep impressions that cause anxiety, and if left to fester can boil over into even stronger emotional outcomes that rob your staff of the energy, heart, and spirit to move forward. The underlying issue is that what the mind can call up, believe, and then react to becomes the individuals' reality. And any display by your staff of a negative, fretful mind-set when under competitive pressure is not what you are looking for.

In addition to competitive actions, other triggers that shape psychological impressions include failed performance of a new product, reduced profits, changing customer behavior, or lost market share.

Such states of mind can initiate another set of emotions; some are accurate, others distorted through false or exaggerated interpretations of what they observe or heard. For whatever the cause, the effect can deepen and degrade into still lower levels of morale. This is especially troublesome when such conditions jeopardize your ability to take positive action.

Here, again, the "supreme requirements of (leadership) are a clear perception ... and an ability to examine the human factors." Your aim, then, is to watch the various human factors and attempt to clear them up through proper communication and involving your people in solutions.

> The man of action must at times trust in the sensitive instinct of judgment, derived from his native intelligence and developed through reflection, which almost unconsciously hits on the right course.
>
> **Clausewitz**

To "trust in the sensitive instinct of judgment" requires *discipline, training, leadership, ambition, and self-confidence* on your part as well as those you manage. We now look more closely at each influence and its impact on leadership.

Discipline

Preparing your staff to bear up under intense competitive conditions involves a good deal of leadership skill. That is, any form of restraint entails sensitivity on your part if you intend to change attitudes and forms of behavior. This is particularly essential when individuals are naturally predisposed to back away from the realities of an uncertain and erratic competitive threat.

Therefore, your aim is to motivate individuals to make audacious efforts to reverse a situation and turn potential defeat into victory. As quoted elsewhere in this book, Clausewitz's sage comment holds true, "in war [confrontation] the result is never final."

Accordingly, your intention is to be totally aware of what psychological effects competitive conditions have on your employees. With that concept in mind, your central task is to inspire individuals by reinforcing the idea that there are always actions that can change dire conditions into successful outcomes.

The central point to understand is that, if handled skillfully, you can turn around a competitive threat and make it an opportunity. In other words, if your counteraction is set off with a disciplined strategy for survival, followed by a strategy plan for growth, you can reignite your staff's creativity and energy to find solutions to a supposed untenable situation. As a parallel issue, taking such an approach goes a long way in strengthening how you are viewed by your subordinates.

For example, one North Carolina-based company that markets organic fertilizer and similar environmentally friendly products reported a grass-roots change in behavior among its employees. When informed by the company president that market conditions were tough and there were few funds for marketing, individuals on their own initiative became their own marketing force. They went on the road and visited retailers to check on displays, talked up their products, and chatted with customers. Instead of taking a plane, staying in a hotel, and charging meals, several people drove their cars more than 1,000 miles each, going so far as to sleep in their back seats.

All ... action is permeated by intelligent forces and their effects.

Clausewitz

Training

What follows from discipline is the essential need for ongoing training so that you can lead "intelligent forces" to meet the demands of a variety of conditions. That means equipping your staff with new skills or updating existing ones to help them cope with such circumstances as the demands of extra travel, long hours of work, or budgetary austerity. Your aim during hard times is to prevent them from caving in to pressures. Instead, you want them to come forward with rational decisions to deal with severe competitive conditions.

What exactly do you want the training to accomplish? Use the following guidelines and modify them for your particular use. First and foremost, within the framework of this book and the nature of a competitive global economy, assume that a competitor will vigorously confront your aggressive moves in ways that would jeopardize your market position. Accordingly, develop specific training sessions aimed at creating a state of readiness. The key word is *readiness: readiness* to take on an opponent going after your company and your market.

Train individuals to develop a competent business strategy plan with a customer-driven outlook and one that leads to a comparative advantage against competition. Components of that training should include making certain that individuals internalize the strategic direction of the plan, they know how to write objectives that complement the vision statement, they prepare strategies to achieve the objectives, and they propose products and services to drive the organization forward.

Train your people in the vital importance of maintaining ongoing competitor intelligence (CI). This is the key component to developing and implementing competitive strategies. The central idea behind CI is that if you know your rival's plans, you are able to monitor his moves. Then you can figure out which strategies will likely succeed and avoid those with little chance of realization. You thereby gain the advantage of knowing where your competitor's strengths are formidable and where he is weak. What better way is there for you to establish status among your people, as you motivate them to deal with potentially harmful situations?

Initiate training that gives your group a fighting edge. As part of that training, the team should learn to deal with input from a variety of viewpoints and sort the information into a coherent and actionable format. (See Chapter 5, "Functions and Responsibilities of a Cross-Functional Team.")

Establish interactive communications within your organization that encourage innovation among all levels of personnel. That means using the power of positive words, encouraging optimistic attitudes, sharing forward-looking plans, and fostering full participation.

There is a need for clear, organized communication that makes available to various groups current information about the following: long- and short-term company and group objectives; market, industry, and technology trends; and intelligence about competitors. Such summaries also include tactical details that concern marketing, customer service, and logistics.

Clear communication, therefore, requires an organization that utilizes a systematized information delivery process, so that there is an easy flow of information up and down the organization. Such systems are readily available; they just need to be put into service (or upgraded.)

What is essential in the temperament of a general is steadiness.

Sun Tzu

One of the overall outcomes of training is to create an environment of trust and understanding, whereby subordinates are encouraged to seize the initiative and act with a sense of purpose and loyalty. As for loyalty, it is a two-edged sword in that it is no giant leap for employees to think their leader is disloyal to the organization, to believe their leader will be unfaithful to them as well.

Also, training should aim toward *innovation*. As applied to readiness, it focuses on truly imaginative and differentiated products or processes, as well as those categories of perceived new products described in Table 6.2. This type of training may require a new level of thought consistent with a Web-based economy. In particular, it requires a company culture that encourages innovation and a leader with a "temperament of ... steadiness."

> When people discuss a general, they always pay attention to his courage.
>
> **Sun Tzu**

Ambition

One of the essential qualities of a leader is ambition. There never was an outstanding leader without ambition. It is the mainspring of all actions. But for pragmatic meaning, ambition must be worthy of the organization's mission and not a pathway to solitary power.

Ambition is difficult to separate from courage. In analyzing great leaders it is generally impossible to decide which of their actions in the face of severe problems bore the mark of courage or that of ambition. Both are characteristics of the truly outstanding leader.

It is constructive ambition and the intense desire to excel that stimulates drive in others. The magic of winning always arouses determination, which gives momentum to the organization. Therefore, nurturing positive ambition is another prime duty of the leader.

Yet the unwelcome reality exists that unrestrained personal ambition does live on, with all its excesses and potentially harmful outcomes. It is uncontrolled raw ambition that destroys employees' careers and the economic livelihoods of communities in which the organizations operate. Such excesses and scandalous executive behavior, which decimated some of the loftiest organizations, became a shocking reality during and beyond the 2008 recession.

Consequently, if you seek competence in leadership, understand how the power of corporate culture interfaces with ambition. Corporate culture, as discussed in Chapter 5, is the cement that binds together all the qualities, gives an organization a unique personality, and forms the underpinnings of unity.

For a general unable to estimate his capabilities or comprehend the arts of expediency and flexibility when faced with the opportunity to engage the [competition] will advance in a stumbling and hesitant manner, looking anxiously first to his right and then to his left, and be unable to produce a plan. Credulous, he will place confidence in unreliable reports, believing at one moment this and at another that.

Sun Tzu

Self-Confidence

As referred to earlier, embedded in excellent leadership is the enormous power of self-confidence, whereby you will perform correctly in a critical situation. Self-confidence comes from the inner feeling that you are competent in your job and will not "advance in a stumbling and hesitant manner."

And what do your employees absolutely desire? They want a self-confident leader who can accurately assess conditions, know what needs to be done, and demonstrate a capability to take appropriate actions.

That means, for instance, when you go into a local marketplace, be watchful that you do not become sidetracked by an isolated incident. Instead, look at the event and fit it into the framework of the entire market scene. Then test it against the broader business objectives of your business plan.

And should your self-confidence still falter and you feel your problems weighing on you more heavily, it is your obligation to take whatever steps you can to strengthen your own determination, rekindle enthusiasm, and most importantly, instill discipline in yourself and those you manage.

Begin by focusing on the strategic direction and the business plan as a whole. Once again, it is vitally important that you retain a broad view of the whole situation. Understand, too, that effective leadership supported by well-thought-out strategies can work as a counterbalance to transitory emotions. This will also help you remain steadfast about implementing the business plan.

Thus, those skilled at making the enemy move do so by creating a situation to which he must conform. Therefore, a skilled commander seeks victory from the situation and does not demand it of his subordinates. Experts in war depend especially on opportunity and expediency. They do not place the burden of accomplishment on their men alone.

Sun Tzu

LEADERSHIP IN THE COMPETITIVE WORLD

Leadership is about responsibility, accountability, and achieving objectives. Leaders inspire their people, organize actions, develop strategies, and respond to market and competitive uncertainty with speed and effectiveness.

Above all, leaders act to win: to win customers, to win market share, to win a long-term profitable position in a marketplace, and to win a competitive encounter before a rival can do excessive harm. If they lose, their organizations and those they manage suffer.

A leader "seeks victory from the situation." That means influencing employees by providing purpose, direction, and motivation, so that they do not "place the burden of accomplishment on their men alone."

Consequently, anyone who influences and motivates people to action, or affects their thinking and decision making, is a leader. Leadership is not only a function of *position*; it is also a function of an individual's *role* in the organization. Consider the following examples where leadership is a combination of position and role.

Verizon Wireless launched its version of a music download service to compete with Apple Computer's highly successful iPod® and iTunes® combination. It took a solid dose of boldness to go against an extraordinarily famous product.

It also took the leadership of numerous individuals in varying organizational positions and roles to link vision, purpose, direction, and motivation to their respective staffs. So that, throughout the organization, Verizon's product designers, engineers, manufacturing, marketing, finance, and an assortment of senior and mid-level managers displayed their allegiance and dedication to activate the corporate vision.

As this multilayered blend of leadership looked to the future, they also worked at the immediate demands of the job. Even with diverse managerial techniques, they demonstrated a common ability to make the objectives clear to those they supervised, which was followed by a vigorous execution of their business plans.

For Verizon to successfully introduce its new product required flexibility and a mix of leadership styles as different situations arose. Further, no one could be cast as the single leader; each also behaved as a subordinate. And all members of the organization worked as part of a team.

What lessons can you take away from the Verizon case? In a market-driven, highly competitive environment, work at developing a personalized, yet flexible managerial style. Anything else will come across to your personnel as artificial and insincere. This is especially important if you expect them to support the organization's overall vision and goals.

Also, if you rely on only one leadership approach, you suffer the consequences of being rigid and will likely experience difficulty operating in situations where a single style simply does not work. Some projects are complex and require different management skills at each stage of development.

For instance, projects in the early stages of development, where creative insight and patient testing for performance dominate, require a far different leadership style from that of pumping up a sales force when launching a new product.

Similarly, products at various stages of their life cycles—introduction, growth, maturity, decline, and phase-out—involve different leadership methods that correspond to the varying market and competitive conditions at each stage.

For those reasons, there is no single leadership style for all occasions. Therefore, model your style to fit your organization's overall objectives. Be certain, too, that it conforms to the individual tasks to be performed by role and function.

There is still another issue that affects leadership: how your personnel feel about the climate within your organization. Climate relates to your employees' perceptions and attitudes about the day-to-day functioning of the organization and their respective units.

Climate, in turn, is allied with corporate culture. Therefore, if you are to develop realistic strategies and implement them successfully, it is in your best interest to define the environment in which you work.

Accordingly, answer the following questions to determine your organization's climate and your role in it. Although you may not be in a position to make changes, at least you can point to the negatives and positives that exist in the everyday workings of the organization and thereby personalize your own leadership style.

Are priorities and objectives clearly stated and do your personnel generally accept them?

Is there a system of recognition, rewards, and reprimands? Does it work?

Do you seek input from subordinates? And do you act on the feedback provided? In particular, do you keep your people informed?

In the absence of instructions, do individuals reporting to you have authority to make decisions that are consistent with your objectives? Do they take the initiative and act in times of opportunity or emergency?

Are there signs of excessive tensions among employees or acts of competitive in-fighting in the organization? What are the causes?

Is your leadership style consistent with your company's values? Is there a working climate of trust? Do other leaders make positive or negative role models?

The following examples illustrate these points.

Google creates a working climate in which its managers display the outstanding qualities of leadership by motivating their employees to innovate in all aspects of their jobs. Recognizing that inventiveness and innovation are the drivers of organizational success, leadership is dedicated to creating a working culture that encourages fresh ideas.

For instance, Google management gives all engineers one day a week to develop their own pet projects, no matter how far from the company's central mission. If work deadlines get in the way of those free days for as much as a few weeks, they accumulate. Also, the system is so pervasive that anyone at Google can post thoughts about new technologies or businesses on an *ideas* mailing list, available company-wide for inspection and input.

What are the leadership traits that support such behavior? First, respect for the individual forms the basis of Google's leadership. In practice it means recognizing and appreciating the inherent dignity and worth of people. And even where some individuals' ideas will not succeed, their efforts are recognized and respected. This is especially relevant working with culturally diverse personnel with a wide range of ethnic and religious backgrounds.

Second, at each level, leaders stand aside and let subordinates do their jobs. They empower their people, give them tasks, delegate the necessary authority, and let them do the work.

The fundamental issue here is that the organization is not going to stop functioning because one leader steps aside. Therefore, central to the job of good leaders is helping subordinates grow and succeed by teaching, coaching, and counseling.

It is the business of a (*leader*) to be serene and inscrutable, impartial, and self-controlled.

If serene he is not vexed; if inscrutable, unfathomable; if upright, not improper; if self-controlled, not confused.

Sun Tzu

BorgWarner specializes in technologies related to fuel economy, vehicle emissions, and stability. It is also an organization that displays effective leadership and operates in tune with the best attributes of a successful twenty-first-century enterprise.

For management, the principal challenge for BorgWarner is to inspire employees to step up the pace of innovation. The challenge is not just hype. A fund of $10 million is set aside as seed money for ideas that are solicited at company innovation summits.

Those events usually result in three or four research projects going forward a year, out of hundreds of ideas submitted. In terms of outlay, the investment represents a mere fraction of what BorgWarner spends on research and development.

Recognizing that technology is a leader's ongoing responsibility, management continually learns how to administer it. Therefore, a central role of leadership at BorgWarner takes the route of harnessing technology to maintain a competitive advantage. Associated with this duo of technology and leadership is the value of a disciplined, cohesive organization that is able to ride out tough times and emerge better than when it started.

Riding the waves of uncertainty also means battling the human tendency of procrastination and the temptation to wait for every scrap of information before making a decision to move forward. Your best approach is to act proactively by developing a business plan with the following general guidelines:

First, establish primary as well as secondary objectives, so that if your main objectives cannot be achieved, you have a fallback position.

Second, develop corresponding strategies and tactics for each of your objectives.

Third, set up monitoring systems to flag problems, which will permit you to manage unexpected situations with contingency plans.

Fourth, work out an exit plan to manage situations that could develop into untenable conditions.

A good plan violently executed now is better than a perfect plan next week.

General George S. Patton, Jr.

By accepting Patton's pragmatic wisdom that decisiveness beats indecision, you can more readily maneuver by indirect strategy (Chapter 1), act with speed (Chapter 2), and be able to move to the offensive (see ahead in Chapter 8). Taking prompt action, however, does not mean impetuous moves whereby you run off with incomplete plans or launch flawed products.

What lessons in leadership emerge from the above examples?

You can achieve excellence as a leader when your people are disciplined and committed to the organization's cultural values. Excellence in leadership, however, does not mean perfection. On the contrary, an excellent leader allows subordinates room to learn from their mistakes as well as successes.

For instance, in the event you have to take over a troubled organization or business unit, you would probably sort through nebulous facts and try to make sense of questionable data. In the process, you are likely to lean heavily on trusting your knowledge, experience, and best judgment to size up the situation quickly, determine priorities, and above all, act. And notwithstanding those systematic and logical moves, you are ready to feed off your gut-level intuition to arrive at final decisions.

Another lesson emerges: As you consider alternative courses of action, you must take into consideration the consequences of each move. Those considerations most often include the ability to assess subordinates and peers for strengths, weaknesses, and a willingness to take action.

Embedded in all those lessons is the self-confidence that you will perform correctly in a tough situation. As discussed earlier, self-confidence comes from the inner knowledge that you are competent in your job.

When using troops, one must take advantage of the situation exactly as if he were setting a ball in motion on a steep slope. The force applied is minute but the results are enormous.

Sun Tzu

USING LEADERSHIP AS A FORCE MULTIPLIER

In addition to the above qualities, leadership demands calmness and patience. In particular, those traits take into account not only how you personally react in a variety of competitive situations, but also observing the behavior of your employees, especially when they are pushed against rivals.

In its fullest sense, leadership considers people as a force multiplier to "take advantage of the situation." Such awareness is vital and will impact your decisions, especially when faced with the critical choice between taking the most daring or the most cautious pathway.

The essential point is that the nature of modern-day business thinking in a competitive arena requires decisive and bold action. And to take bold action entails effective leadership with motivated and disciplined employees ready and willing to implement the strategy plan, as if you "were setting a ball in motion on a steep slope."

Therefore, develop your own personal style of leadership. That is, leadership is not a one-size-fits-all style. Tailor your approach to your character and personality. However, the key here is to be certain that your style harmonizes with your organization's prevailing culture—and, of course, with the job at hand.

Also, while leadership relies on basic tenets regarding the treatment of individuals and dedication to clear-cut objectives, it is not an off-the-shelf package created by someone else for you to adopt. It should vary with the strategy you develop and the type of workforce you manage. Accordingly, look closely at your own behavior and match it to the attributes mentioned in this chapter.

If your strategy is going to succeed, your subordinates are the ones who must convert it to action. Therefore, take note of your employees' behavior and look out for any of these ominous signs: *unrest, grievances, uncommunicativeness, uncertainty, and any signs of agitation*—before, during, and after a competitive encounter.

Notice, too, if their feelings are anchored to the fear of losing pride, status, employment, as well as any other areas of friction. As important, look for any psychological effects, for instance, individuals visibly showing emotions, such as losing face in the eyes of their peers.

> If the skill of a general is one of the surest elements of victory, it will be readily seen that the judicious selection of [*managers*] is one of the most … essential parts of the policy of state.
>
> **Jomini**

Finally, the following attributes will get you closer to creating a force multiplier effect by projecting an optimistic leadership style.

Interpersonal Skills

These affect your dealings with people. Skills include coaching, counseling, motivating, and empowering. It is the ability to communicate your intent effectively, without impatience or anger. If you want your subordinates to be calm and rational under pressure, you must set the tone.

Conceptual Skills

These enable you to envision business-building strategies. Such skills require sound judgment, as well as the ability to think creatively, reason logically, and act ethically. It also means sensitivity to the group's shared set of beliefs, values, and assumptions about what is important.

Technical Skills

These are job-related skills you must possess to accomplish all your assigned tasks and functions. Included here is a wide-ranging awareness consisting of (1) respect for the diverse backgrounds of your personnel, (2) responsiveness to the varied cultures of your markets, (3) sensitivity to the customs and traditions of the intermediaries and customers with whom you interact, and (4) insights into the technical skills and technology practices of your competitors.

Tactical Skills

These apply to solving tactical problems, which include handling localized situations concerning the deployment of resources. Tactical skills combine with the above interpersonal, conceptual, and technical skills to achieve objectives.

In the next chapter we get into the details of joining your leadership capabilities to creating a morale advantage.

7

Engage Heart, Mind, and Spirit: Create a Morale Advantage

The final deciding factor of all engagements ... is the morale of the opposing forces. Better weapons, better food, and superiority in numbers will influence morale, but it is a sheer determination to win, by whomever or whatever inspired, that counts in the end.

Study men and their morale always.

Field Marshal Wavell

A vital responsibility of any leader is to heighten employee morale.* Doing so supports unity within the group and taps the inner strengths of individuals. As worthwhile, it prevents them from faltering when stress increases due to harsh market conditions. Morale also tends to energize individuals with the resolve to act decisively under competitive pressures, which, at critical moments, could be the "final deciding factor of all engagements."

As a dominant form of behavioral expression, morale affects day-to-day employee performance and contributes ultimately to how well a business plan is implemented. Consequently, morale, in its most pragmatic form, functions as a gauge of how people feel about themselves. In particular, it shows the extent to which individuals will participate in a team effort and the confidence they show in their leaders.

Morale, then, is shaped by common values, such as loyalty to fellow workers and a belief that the organization will care for them. Where morale

---

* Morale intertwines with other influences, such as personality, creativity, experience, and intuition. It also includes the emotions, culture, and psychology of individuals, as well as the influences of training and motivation developed within the organization.

is at a high level, it results in a cohesive team effort "with a sheer determination to win."

> The appropriate season is not as important as the advantages of the ground; these are not as important as harmonious human relations.
>
> **Sun Tzu**

Successful leaders know that in the face of the inevitable problems and reversals, morale holds a group together and keeps it going. They tend to use a variety of approaches to create "harmonious human relationships." Most are offshoots of two dominant methods: the popular, *participative*, and touchy-feely style that many executives embrace today, which is diametrically the opposite of the *autocratic* style that is less prevalent among major organizations, but still finds a home in some top firms.

A noteworthy example where the two dominant approaches lived successfully (at different times) is *Home Depot*. From its beginnings in 1979, store managers enjoyed immense autonomy to make their own decisions and run their own operations. Founders Marcus and Blank had installed a decentralized, entrepreneurial business model and used a highly personalized leadership style. While the work was demanding, the company grew in a low-profile, collaborative, and mutually respectful working climate. With its humanistic managerial and leadership methods, the Home Depot chain expanded to become one of the largest retailers.

A turning point came in 2000 with a change in management, which began a managerial and cultural transformation. Compared to its cultural heritage, the managerial changes were venturesome and audacious.

Built on a military organizational model, the then- CEO imported ideas, people, and concepts from the armed forces. Overall, some 13 percent of Home Depot's employees had military experience, versus 4 percent at Wal-Mart.

Sweeping moves were initiated to reshape Home Depot into a more centralized organization with a command-and-control management structure. Every major decision and goal at Home Depot flowed down from headquarters.

The approaches rested on building a disciplined corps that would be predisposed to taking orders, could operate in high-pressure environments, and execute strategies with high standards. The spirited methods rekindled stellar financial performance during the CEO's tenure. At its highest point, sales soared, growth rose at a double-digit rate, and profits more than doubled.

Which leadership style is better? Should either be considered a consistent style for all seasons? What benchmarks should be used to determine style: customer satisfaction, degree of competitive aggressiveness, employee performance, stock price (if applicable), gross margins and profits, or high levels of innovation and inventiveness shown by employees?

How about the working climate? Should there be a hardball working climate? Should the surroundings match the innate personality of the leader? Or should it be a flexible work environment that correlates with competitive conditions?

Also, what value would you place on employees' attitudes? To what extent do they impact morale, spirit, and the seeming intangible of *heart* on the outcome of your business plan?

What steps should you take to adopt either of the two leadership approaches? What commonalities exist between the two styles—if any?

> Loss of moral equilibrium must not be underestimated merely because it has no absolute value and does not always show up in the final balance. It can attain such massive proportions that it overpowers everything by its irresistible force. For this reason, it may in itself become a main objective of the action.
>
> **Clausewitz**

MOTIVATIONAL BEHAVIOR

For you to find personal answers to the above questions, and to internalize the contrasting management styles of Home Depot, you will find it highly useful to first familiarize yourself (or get reacquainted) with the gurus of motivational behavior, namely, *Frederick Herzberg, Douglas McGregor, Abraham Maslow,* and *William Ouchi.* They spawned motivational theories in the twentieth century that are still valid today, some of which parallel Clausewitz's viewpoint that achieving "moral equilibrium ... overpowers everything by its irresistible force." Using these ideas to elevate the morale of your people will likely have a positive impact on implementing your business plans.

Herzberg's Motivation–Hygiene Theory

The most important element of Herzberg's theory is that the main motivating factors are embedded in the satisfaction gained from the job itself. He reasoned that to motivate an individual, a job must be challenging, with sufficient scope for enrichment and interest. Motivators—often called satisfiers—are directly concerned with the satisfactions gained from a job.

In contrast, a lack of motivators leads to over-concentration on what Herzberg called hygiene factors—or dissatisfiers—which form the basis for complaints. Table 7.1 presents the top six factors causing dissatisfaction and satisfaction.

> Pay heed to nourishing the troops; do not unnecessarily fatigue them. Unite them in spirit; conserve their strength. Make unfathomable plans.
>
> **Sun Tzu**

McGregor's XY Theory

This theory remains a valid and predictable means by which you can develop a positive leadership style "to unite them in spirit." It is central to organizational development and corporate culture.

McGregor maintained that there are two fundamental approaches to managing people, popularized as Theory X and Theory Y. Theory X tends to use an authoritarian leadership style. In contrast, Theory Y leans toward a participative approach. In today's environment, Theory Y generally is accepted as producing better performance, in that it allows people more

TABLE 7.1

Herzberg's Factors Affecting Motivation

Factors Leading to Dissatisfaction	Factors Leading to Satisfaction
1. Organizational policy	1. Sense of achievement
2. Quality of management	2. Level of recognition
3. Relationship with boss	3. Intrinsic value of the job
4. Working conditions	4. Level of responsibility
5. Wages	5. Opportunities for advancement
6. Interpersonal relationships	6. Status provided

TABLE 7.2

Contrasting Views of Theory X and Theory Y

Theory X (Authoritarian Management Style)	Theory Y (Participative Management Style)
The average person dislikes work and will avoid it, if possible. People must be forced to work toward organizational objectives. Individuals prefer to be directed, look to avoid responsibility, generally lack ambition, and want security above all else.	Effort in work is natural and enjoyable. People will apply self-control and self-direction in the pursuit of organizational objectives, without external control, or the threat of punishment. Commitment to objectives is a function of rewards associated with their achievement. People usually accept and often seek responsibility. People use a high degree of imagination, ingenuity, and creativity to solve business problems.

latitude to grow and develop in self-motivating surroundings. The two styles are contrasted in Table 7.2.

> Throw the troops into a position from which there is no escape and even when faced with death they will not flee. For if prepared to die, what can they not achieve?
>
> Then officers and men together put forth their utmost efforts. In a desperate situation they fear nothing; when there is no way out they stand firm … they are bound together.
>
> **Sun Tzu**

Maslow's Hierarchy of Needs

Maslow viewed people as basically trustworthy, self-protecting, self-governing, and, when put to the test, able to come "together [and] put forth their utmost efforts." Further, he believed that individuals' innate tendencies are toward growth.

According to Maslow's theory, there are five types of needs that must be satisfied before a person can act unselfishly. Table 7.3 shows these needs, which are arranged in hierarchical order and usually shown as a pyramid. The path is to satisfy one set of needs at a time, beginning with the physiological need, then move upward to self-actualization.

TABLE 7.3

Maslow's Hierarchy of Needs

Physiological needs

These cover the basic functions of comfort and maintenance of the body, such as food, drink, heat, shelter, sleep, and health.

Safety needs

These refer not just to physical safety and protection from harm, but also to such areas as financial security, employment, medical and legal assistance, and all means that maintain stability.

Belonging needs

These indicate the need for human contact: family, friends, relationships, teams, and general contact in society.

Esteem needs

These recognize the need for status, power, prestige, acknowledgment, respect, and responsibility. Such requirements provide individuals with a higher position within a group.

Self-actualization needs

After all the previous needs have been satisfied, the top of the pyramid deals with the individual's need to reach for his or her full potential and strive for individual destiny.

> By moral influence I mean that which causes the people to be in harmony with their leaders, so that they will accompany them in life and unto death without fear of mortal peril.
>
> **Sun Tzu**

Ouchi's Theory Z

This theory is essentially a combination of all that is best about McGregor's Theory Y and modern Japanese management. It places a large amount of freedom and trust in "people to be in harmony with their leaders." It also assumes that they possess strong loyalty and interest in teamwork and in the organization.

Theory Z places a great deal of reliance on the attitudes and responsibilities of workers, whereas McGregor's XY theory is mainly focused on management and motivation from the manager's and organization's perspective.

Now, what conclusions can you come to from the earlier questions related to the Home Depot case and the review of classic motivational theories? There is sufficient evidence to draw two immediate conclusions:

First, it is possible for both autocratic and participative managerial styles to produce positive results, depending on the level of morale among the participating employees, and the leadership skill of the manager. Second, even in a somewhat autocratic, quasi-militaristic work climate, where there is a defined mission, a set of clear and measurable objectives (even when dictated by the CEO), understandable communications, and clearly stated expectations, this style can flourish.

Consequently, as pointed out in Chapter 6 on leadership: "If you rely on only one leadership style, you suffer the consequences of being rigid and will likely experience difficulty operating in situations where a single style simply doesn't work."

What follows is a set of conditions common to whichever style you adopt:

Hold fast to the definitive object of all business, which according to the late management scholar, Peter Drucker, is to *create a customer.*

Remove obstacles that deprive people of the ability to gain pride in quality of service and delivering innovative products.

Break down inhibiting barriers among diverse staff and create collaborative cross-functional teams.

Introduce a work environment in which the emphasis is on managers leading, not merely supervising.

Commit to long-term goals, such as attaining market leadership, developing leading-edge products, or maintaining superior service and product quality.

Reduce the use of fear as a motivator.

Encourage employees to express ideas; then listen to them and respond.

Promote self-improvement as an ongoing imperative. Institute continuing employee training and education to advance their skills, personal growth, and chances for career advancement.

He whose ranks are united in purpose will be victorious.

Sun Tzu

Another conclusion that can be drawn is that morale creates cohesion or being "united in purpose." Cohesion allows you to tap more readily into the inner strengths of individuals, thus eliciting the resolve and determination to act decisively under competitive pressures. As important, it prevents them from faltering when stress increases due to a tough market situation.

> When ignorant that the army should not advance, to order an advance or ignorant that it should not retire, to order a retirement: This is described as hobbling the army.
>
> **Sun Tzu**

There is another reality you have to face about morale: Some managers never seem to worry about morale. No obstacles seem to bother them. *Here's what I want you to do; now do it* is a familiar command. Most often, however, they experience incredible disorder in the follow-on moves, which could cause a "hobbling" effect. They are the great leaders for a day, until the moment that some negative outcome overwhelms them.

The greater and more reliable reality is that marketplace conflicts rely on positive outcomes through individual fortitude, triggered by the potent force of morale. And that takes ongoing training, excellent leadership, and intimate concern for employees' well-being. These are the factors discussed in this and the previous chapter.

> Heart is that by which the general masters. Now order and confusion, bravery and cowardice, are qualities dominated by the heart.
>
> Therefore the expert at controlling his enemy frustrates him and then moves against him. He aggravates him to confuse him and harasses him to make him fearful. Thus he robs his enemy of his heart and of his ability to plan.
>
> **Sun Tzu**

MORALE LINKS TO THE HUMAN HEART

Central to morale and the starting point in all matters pertaining to competitive encounters are "qualities dominated by the heart." Heart collectively describes the emotional qualities by which you lead and the means by which you reach individuals. These encompass unity, camaraderie, purpose, duty, and hope, including the principal factors covered in Maslow's Hierarchy of Needs.

In day-to-day organizational life, emotions materialize in conflicting forms. These are expressed in the extremes of order or confusion, commitment or indifference, boldness or fear, loyalty or deceitfulness. They are the realities that typically dominate the heart.

Not only does heart underlie your role as a manager, it also reflects in your outward behavior and ability to perform as an inspiring leader. That is, your demeanor and attitude filter down and impact the performance of staff members who might be low on morale and stripped of courage.

Accordingly, you have to show confidence and display the discipline and grit that will prevent you from caving in at every obstacle. Heart is how you create stability out of confusion, enthusiasm from discouragement, and bravery out of cowardice. These, too, are qualities that dominate the heart.

On the other hand, if you personally lose courage, the winning spirit, and decisiveness to complete the planned efforts, then your ability to lead under stressful situations is at serious risk.

> Therefore those skilled in war avoid the enemy when his spirit is keen and attack him when it is sluggish. This is control of the morale factor.
>
> **Sun Tzu**

Reaching the heart and mind applies equally to competitors. Strategically, your aim is to frustrate the competing manager and "attack him when (his spirit) is sluggish." If you aggravate and confuse your competitor into making hasty and unwise decisions, you rob him of the heart to plan with the winning spirit and courage appropriate to the demands of an uncompromising marketplace.

As a result, in the hands of an astute manager, this psychological aspect of human behavior becomes an effective strategy tool to unbalance a competitor. Heart, therefore, contains the highly humanistic components that profoundly impact your capacity to manage your people, your competitors, and your ability to implement a business plan.

Tactically, if you move into unfamiliar markets where competitive, economic, and negative consumer influences loom as unexpected barriers, then it is only through the spirited efforts of your people that you can overcome obstacles and push forward. While the inner stirrings of fear and uncertainty can fretfully tug against you, there are also the dominating effects of heart, with all their remarkable qualities, that in the end determines who wins or loses.

In the final analysis, successful performance in a competitive encounter is a matter of morale and reaching the hearts of your people. In all matters that pertain to an organization, it is the human heart that reigns supreme during times of conflict. Careless managers rarely take it into account, resulting in irreversible and unfortunate outcomes.

> The appreciation and understanding of moral factors can only be perceived by the inner eye, which differs in each person, and is often different in the same person at different times.
>
> **Clausewitz**

MORALE LINKS WITH TRUST

Decisive action builds trust and supports morale. So, too, do the manager's "appreciation and understanding" words of encouragement and the projected image of reliability, integrity, and competence. These expressions should be activated by well-intentioned motives that serve a common interest instead of being self-serving.

Trust, however, can be undermined by the sometimes explosive behavior of subordinates, which is due to the impulsive nature of people. They nervously shrink back and find danger in any effort in which they do not foresee the possibility of success. Personnel also tend not to act as passive obedient instruments, but as very anxious and restless individuals who wish to finish things quickly and know in advance where they are going.

The purpose of focusing on trust, therefore, is to overcome those negative feelings among individuals who, because of various life experiences, mistrust and doubt. Some have valid reasons due to having been misinformed and misled in the past. Nonetheless, it is your challenge to create trust and convince them that you have their best interest at heart.

> A skilled commander seeks victory from the situation and does not demand it from his subordinates.
>
> **Sun Tzu**

MORALE LEADS TO UNITY

It is your responsibility to "seek victory from the situation" and not to "demand it from [your] subordinates." Your staff must know that support exists at the highest levels of the organization. Also, they must see and feel your physical presence. And they need to absorb the psychological comfort and confidence that would sustain their morale and motivate them to keep trying, even under adverse conditions (see Table 7.1).

As for unity of effort, it is not isolated actions that make high-performing employees. Rather, it is the collective efforts of individuals within teams interacting among themselves, as well as with those in dissimilar teams, that cultivate unity. (See the Johnson & Johnson case below.)

For that reason, it is your essential role to encourage such interaction. Once again, this is the purpose of a team. In the end, it is the manager who knows his or her people best and understands the reasons behind such highly charged displays of behavior as anguish and fear or courage and dogged determination.

Therefore, you cannot be a stranger to your employees; it is your influence and physical presence that affects morale. If they feel themselves no longer supported, it creates an untenable situation for you.

Unity also requires that you sustain confidence in your personal ability to lead. If, on the other hand, you are habitually gripped by fear, then you should reconsider your ability to lead your group.

Unity is also nurtured by a cohesive corporate culture, which is reinforced by an ethical climate built on core values, as illustrated below in the Boeing case. A healthy culture fortifies your leadership role and helps guide your organization or business unit's strategy. Some ways in which morale impacts business strategy are

- Morale suffers and defeats become competitive disasters without sufficient training of leaders and staff.
- Morale fades when there is no inspiring goal to fight for.
- Morale weakens when employees work in a state of uncertainty.
- Poor communication affects morale, with a corresponding serious effect on implementing strategies.
- Lack of commitment to a competitive strategy with a determination to win affects morale.
- Unethical behavior impacts morale. (See the Boeing case below.)
- An inability to create a long-term vision and develop attainable objectives influences morale.
- Undisciplined behavior that readily gives in to fear sways morale.
- A leader who procrastinates and visibly displays an inability to handle responsibility has a bearing on morale.
- A leader who stifles employees' input has a negative influence on morale.

- A leader's inability to accurately assess a competitive situation shapes morale.
- A leader who does not know how to reach the heart, mind, and spirit of a group and instead shows more interest in his personal agenda impacts morale.

Consequently, leaders seek to shape their companies' or business units' cultures to support their visions, articulate their goals, and improve their overall performances through high morale.

MORALE INTERFACES WITH INNOVATION

Innovation is the new differentiator. Therefore, use all your efforts to keep jobs stimulating. And when employees' creativity has significant results, express appreciation publicly for jobs well done. Your approach is to be evenhanded, open-minded, and fair.

Further, in all initiatives associated with innovation, support your people, communicate a vision of growth, provide them with the resources to do their jobs, and motivate them to learn and grow. As important, stand by them if something goes awry. Or else you shut them down—along with creativity.

Also work at cultivating your group's mini-culture. 3M safeguards its culture by enlisting old-time employees to recount company history and hand down stories of 3M's long tradition of innovation to new engineers. Expanding the technique throughout the organization, after a while, every new employee is able to recite the foundation precepts that are the underpinnings of the organization.

Of course, be sure that what you focus on with your group complements the overall organizational culture. In doing so, you will find that parts of the culture deal with the work environment and issues related to quality of life. Do not expect innovation and performance from troubled and discouraged individuals with *life* issues.

What follows is for you to establish criteria for evaluating innovation. You need to let the staff know how they will be measured. Frequently used measures of innovation include the following:

- Overall revenue growth that is attributed to innovations
- Percentage of total sales from new products or services

- Level of customer satisfaction
- Improvements in market share
- Comparisons of innovations measured against chief competitors by product category and market segment
- Actual number of new products or services launched
- Rate of new product successes
- Other criteria required by your company

[There is] the so-called theory that weapons decide everything. Our view is opposed to this; we see not only weapons but also people. Weapons are an important factor, but not the decisive factor; it is people not things that are decisive.

Mao Tse-tung

THE RELATIONSHIP BETWEEN MORALE AND TECHNOLOGY

So much has been said about the wonders of technology, to the point of minimizing the human element and, in particular, diminishing its impact on morale. Where such a disparity exists, it is likely to have a serious impact on the organization. Individuals are still the most significant element in conducting business, so that "it is people not things that are decisive" in achieving a favorable outcome.

Even with state-of-the-art technology and a formidable strategy, a plan could languish if employees are not personally motivated with the spirited will to succeed. Stated another way, new technologies can have limited value in the hands of dispirited employees, no matter what their number. It is therefore essential to work at developing employee morale, which is often rooted in the organization's culture.

Boeing Company provides a practical example. The aerospace manufacturer met the enemy and it was itself. Although a high-flyer in technology, research, and technical expertise, it was plagued by a dysfunctional culture laced with ethical violations, bitter infighting among business units, and a wave of questionable business dealings characterized by a win-at-all-costs mentality.

The cumulative effect of those difficulties at one point resulted in Boeing losing its position as the commercial aircraft manufacturer to Airbus.

Also, it was denied access to lucrative government contracts worth billions in potential revenue for several months. Those painful outcomes occurred during the tenure of two past CEOs. A turning point began with the appointment of a new CEO, W. James McNerney, Jr., and his unrelenting moves to revamp Boeing's torrid culture. Recognizing the toxic effects from deep-seated problems, McNerney immediately set out to rebuild Boeing. The major moves included the following:

1. Instilling a culture built on ethical standards of honesty, integrity, and moral behavior. Previously, executives were silent when colleagues used unethical means to secure business contracts. Under the new guidelines, they were encouraged to speak out against such practices. To give authority to the edict, compensation was tied not only to higher performance but also to higher ethical standards.
2. Encouraging constructive communication. Efforts were implemented to stop the practice of hoarding vital information, especially technical data, and denying other business units access to information. Compliance was tied to managers' compensation. The overall aim was to create a common culture and an allegiance to long-term growth.
3. Rewarding managers for meaningful performance. Promotion and increases in compensation were rewards for boosting productivity, as well as for driving growth over a three-year cycle.
4. Using a team approach. Emphasis was on building teams with responsibilities for standardizing and streamlining production lines. The goal was to generate consistent results, lower costs, and improve productivity across all areas of the company, whether on an airplane assembly line or in a research lab. A parallel goal was to end the bitter infighting that prevailed under two previous administrations.

As part of implementing the cultural reformation, McNerney began exerting more effective central control over Boeing's three divisions. Analogous to McGregor's Theories X and Y, he centralized operations and drew a Theory X hard-line approach against unethical practices and demanded fair business dealings.

At the same time, McNerney moved ahead with a healthy infusion of Theory Y by managing with humanistic leadership. He acted attentively to the small things such as recalling individual's names, listening carefully to their presentations, and not embarrassing subordinates in public.

He aimed for a work environment in which people spoke out when they saw something damaging, from sounding off about unethical behavior to stopping a production line if something looked wrong.

> Now there are five circumstances in which victory may be predicted:
> He who knows when he can fight and when he cannot will be victorious.
> He who understands how to use both large and small forces will be victorious.
> He whose ranks are united in purpose will be victorious.
> He who is prudent and lies in wait for an enemy who is not, will be victorious.
> He whose generals are able and not interfered with by the sovereign will be victorious.
>
> **Sun Tzu**

MORALE GENERATES MOMENTUM

Should you be embroiled in an untenable situation, it is in your best interest to maintain momentum and keep the possibility of success on your side. Do not think of your actions as unreasonable or impossible. It is always reasonable and there are always possibilities. As pointed out earlier, Clausewitz observed that in conflicts "the result is never final."

Often, the determining factor for climbing out of a tough predicament and gaining the initiative is to lean heavily on the quality of your personnel. This is possible when you address key issues affecting their morale, such as activating campaigns that lead to growth and creating a feeling of unity within the group. (See the section above, "Morale Leads to Unity," for a list of ways in which morale impacts business strategy). These areas have as their underpinnings spirited leadership, a clearly understood vision or objective, and competent business strategies.

To turn a potential failure into a sure win also depends on continuing training and making certain all the attributes cited above come into play. Blended together, they give individuals confidence to carry on in spite of any inclination to concede defeat. The following example relates to these points and the urgency to maintain momentum.

Johnson & Johnson pushed hard to maintain momentum in one of its important and increasingly competitive medical devices business.

This blue-chip company has had many notable product successes, as well as its share of failures.

At one particular critical period, there was a clear and urgent need to drive product innovation. Without a solid jolt in the form of new products and market expansion, earnings could have suffered and its market leadership would have been threatened.

Managers placed great emphasis on innovation with resolve in a variety of ways: Organizationally, internal start-ups were established. These groups acted as if they were totally independent. Some were housed in unusual locations, such as warehouse-like settings where entrepreneurial inspiration tended to surface more rapidly. If any group came up with a hot idea, they created a business plan and went hunting for financing from J&J's venture-capital arm or from one or more of the firm's existing businesses.

Possibly the most dramatic organizational change was to establish a centralized, cross-functional position for a chief science and technology officer. The change was noteworthy in that the position existed in an already highly decentralized organization where J&J's numerous businesses operated autonomously.

The officer's primary objective was to find innovative ways to harness the creative energies of individuals in various business units and encourage them to search out ways to combine medical devices and drugs in one product. A secondary objective was to spot opportunities in markets that any one of J&J's existing business units might have overlooked.

The new position was also culturally significant in that in a corporate environment that thrived on autonomy, the change was geared to foster cooperation from dissimilar groups and get them to work together and share thinking, technical expertise, and purpose. The initiative was successful in kick-starting the product development effort so that momentum was geared-up for the long haul.

> The effects of physical and psychological factors form the organic whole which, unlike a metal alloy, is inseparable. One might say ... that the physical factors seem little more than the wooden hilt, while the moral factors are the precious metal, the real weapon, the finely honed blade.
>
> **Clausewitz**

What can you learn from the Johnson & Johnson case? First, overall employee behavior is shaped by an organization's culture, which comes to life through its history, ideals, motives, and deep-rooted beliefs. Second, organizations remain inherently at risk if managers neglect to strengthen the morale of their employees. To activate morale and maintain momentum calls for harnessing the creative energies of personnel, in product development, manufacturing, marketing, finance, or any other corporate function. The outward effects of such an effort will show through employees' respect and confidence in their leaders, involvement and desire for team unity, confidence in their fellow workers, and willingness to move forward even against tough barriers.

One of the increasingly used management tools to foster inspired collaboration and manage the informal interactions among diverse teams of employees is *social network analysis*, which maps the interacting networks of groups and individuals. Useful for small or large organizations, the tool has the following advantages:

1. Social network analysis exposes any conspicuous gaps where groups are not interacting but should be.
2. Network maps point out where the frequent creative activities take place and offer concrete information to manage interactions.
3. The analysis shows the invisible, informal connections among people that are missing on a traditional organizational chart.
4. Social network analysis is excellent at providing valuable input about pinpointing individuals who consult with others most often and those they turn to for specific types of expertise.

With the current emphasis on maintaining a momentum of innovation and sustaining morale, if you map the informal networks, you are better able to manage talent and nurture your organization's most expert employees. Such companies as *IBM, Capital Financial Corporation, Procter & Gamble,* and *Merck* use social network analysis to spur innovation, spot talent, and share information among diverse groups that do not normally work together.

In sum, superior organizations and effective leadership "are the physical and psychological factors that form the organic whole," not only in thinking up new and better products, but in reinventing themselves for evolving events.

> One who has few must prepare against the enemy; one who has many makes the enemy prepare against him.
>
> **Sun Tzu**

OBSTACLES TO FOSTERING MORALE

Barriers to building morale will appear. The most significant ones for which you must "prepare against the [competition]" are described below.

Deficiency in Planning Skills

Planning within a team setting can be a unifying activity. Organizations such as *3M,* global IT services provider *HCL Technologies*, Linux provider *Red Hat*, and defense contractor *Rite-Solutions* go beyond teams of limited size and open up the planning process to large numbers of individuals within their organizations.

Planning sessions involve individuals who were previously frozen out of the processes of setting strategic directions, prioritizing objectives, and developing strategies. Pulling in those diverse frontline perspectives made the plans more insightful, pragmatic, and actionable. Another favorable unifying aspect resulted in building enthusiasm by getting those individuals aligned with their particular company's overall strategic direction.

Therefore, it is in your best interest to hone planning skills in others and initiate a collaborative approach. There are numerous formats for developing plans. Most have similar characteristics. Follow any of the popular formats, or use a standardized one that already exists in your firm. (Appendix 3 describes a Strategy Action Plan that you can customize for your company or business unit.)

Within the context of building morale and unity, merely acquiring planning skills is only part of the issue. The more relevant one, as illustrated with the above organizations, is the team's ability to *think* strategically. That means honing skills that permit you and others to look forward at least three to five years.

Further, thinking strategically translates to the effective application of strategy, which is the central purpose of this book. Thus, strengthening planning skills is a means to open your mind to abundant possibilities and to foster the same type of thinking within your team.

Lack of a Strategic Outlook

As pointed out above, this stumbling block to improving morale can be addressed in many instances by actively involving large numbers of your staff in developing a collaborative approach to business plans. The companies that have become leaders in their fields have mastered the process of developing a vision of who they are and who they want to be over the long term. In addition to those already mentioned, other noteworthy companies include *General Electric, Procter & Gamble, Google, Home Depot,* and *Wal-Mart.*

> The musical notes are only five in number but their melodies are so numerous that one cannot hear them all.
> The primary colors are only five in number but their combinations are so infinite that one cannot visualize them all.
>
> **Sun Tzu**

Absence of Creativity

Creativity is generally understood to encompass imagination, innovation, and all those factors that lead to originality, inventiveness, and ingenuity. Where a lack of creativity exists, the central issues again point to leadership, morale, and training. Consequently, the task is to create a work environment whereby groups participate in making "their combinations ... so infinite that one cannot visualize them all."

On one occasion, 3M invited all of its sales, marketing, and research and development employees to a Web-based forum called InnovationLive, which over a two-week period attracted more than 1,200 participants from over 40 countries and generated more than 700 ideas. The end result was the identification of nine new future markets with an aggregate revenue potential in the tens of billions of dollars. Since then, 3M has held several additional InnovationLive events, with more planned for the future.

> Now the method of employing men is to ... give responsibility to each in situations that suit him. Do not charge people to do what they cannot do. Select them and give them responsibilities commensurate with their abilities.
>
> **Sun Tzu**

Inadequate Self-Development

One of the identifiable characteristics of a well-run organization is raising individuals to a level where they have skills to take on "responsibilities commensurate with their abilities." That includes a workable system for ongoing learning, career advancement, and self-development, which heightens morale and cultivates all the ingredients that go into a company's value system.

This barrier aligns with the intrinsic importance of developing your human capital. As Sun Tzu wisely points out: "Do not charge people to do what they cannot do."

Minimal Self-Confidence

Lack of self-confidence shows up as indecisiveness, lack of vision, over-caution, or unwillingness to attack a problem or pursue an opportunity with vigor and boldness. There are too many signs of such negative behavior to hide this human flaw from employees.

Like an affliction, such damaging mannerisms can spread to all those who are exposed to them. If the lack of self-confidence cannot be converted to a positive attitude through self-motivation and outside assistance, the manager should step down before any further damage is done.

Inferior Communication Skills

A major deterrent to maintaining morale is an inability to maintain superior communication capabilities. You should find out which forms of communication maximize interactions. The choices are many, from electronic devices and social media to one-on-one contact.

For instance, to encourage productive communications, *BMW* relocates many of its engineers, designers, and managers to its central research and innovation center to design cars. Face-to-face team contact reduces late-stage conflicts and speeds development times.

> There has never been a protracted war from which a country has benefited.
>
> **Sun Tzu**

Procrastination

One of the sure ways to dampen morale is by unwarranted delays and general patterns of procrastination that leave junior managers and field personnel frustrated (see Chapter 2). Fast-changing consumer demands, global outsourcing, open-source software, near-instant movement of capital across oceans, and electronic transfers of knowledge from the field to decision-making executives at headquarters make procrastination a dangerous managerial trait. It leads to "protracted" campaigns from which you may never benefit.

Consequently, with the pace of innovation increasing exponentially, new technologies spawning new industries and disrupting old ones, and communication networks being created with astonishing speed, delays are a perilous detriment in almost every type of business and competitive scenario.

Volatile Conduct

Of worrisome concern is unpredictable behavior that is inconsistent with market conditions. Employees can understand the need for a flexible managerial style or even a certain amount of eccentricity, if it is understood and accepted as part of a manager's inherent personality. However, they are unable to tolerate inconsistency and sudden erratic displays, particularly if there is no apparent reason for what could be perceived as unacceptable behavior.

Discouragement

If your team lacks drive and is easily discouraged, you can use the action steps that follow to activate a morale advantage. In many instances the morale problem radiates from poor leadership skills, which calls for self-examination to pinpoint weaknesses.

In other cases, remedying the faults may be out of your hands, due possibly to the lack of involvement from the highest levels of management, compensation problems, career advancement concerns, and similar issues. Some of the obstacles to building morale are listed below:

- Employees' lack of trust in the manager's ability to make correct and timely decisions
- Executives' failure to view self-confidence as a powerful motivator to move individuals forward
- Lack of a meaningful effort toward creating group unity through collaborative planning and team building (see the examples described above of companies that successfully built group unity)
- Inadequate communication systems for employees to voice their opinions and discuss product ideas, or the likelihood that ideas may be dismissed without explanation

- Few initiatives for self-improvement or chances for advancement
- A work environment that does not encourage innovative and creative thinking (see the examples above of companies that successfully encouraged these qualities)
- Demotivators, such as job boredom, unfairness, casting blame, lack of recognition, and clamping down on mistakes
- Employees who are doubtful of management's true commitment to keeping up with changes in its industry, as well as the global marketplace

When using troops, one must take advantage of the situation exactly as if he were setting a ball in motion on a steep slope. The force applied is minute but the results are enormous.

Sun Tzu

CREATING A MORALE ADVANTAGE

Sun Tzu refers to using troops (for our purpose, employees) as "setting a ball in motion" and "the force applied is minute but the results are enormous." These metaphors are certainly applicable when considering all that is currently known about the potential forces that contribute to morale. With that aim in mind, use the following list as a guide to creating a morale advantage for your group:

Manage through availability and visibility. Show genuine interest by listening to employees' problems, complaints, and other issues. Fundamental to building morale is demonstrating to your people that you are genuinely interested in them and their growth. That means being visible, sincere, and treating them as valued employees who have insight, experience, and special knowledge. One executive described his visibility technique thus: "Manage by walking around."

Manage with integrity and transparency. Within the range of what is permissible or confidential, explain management's actions to your employees. Provide information and guidance on matters that affect their security. (Review Herzberg's concepts cited above.) Keep your people up to date on all business matters affecting them and tackle unfounded rumors before they do damage.

Develop plans and prioritize actions. With input from others, prioritize those opportunities worth exploiting and rank problems needing solutions. Not only do you maximize resources, time, and energy, but you also involve the team in viewing the big picture and understanding the reasons behind the prioritizing.

Support a collaborative work environment. Create an environment of expectation and excitement, such as the physical surroundings described in the Johnson & Johnson case. It should also be your intention to set up a positive psychological environment that encourages and, as important, anticipates success.

Communicate often and openly. A foundation requirement for activating morale and "setting a ball in motion" is communicating up, down, and across levels. Doing so puts problems and opportunities out in the open and spurs the free flow of ideas to overcome difficulties and seize opportunities. Open communication encourages others to think creatively and bring to the surface any detrimental issues that would hamper a team's progress. There are numerous possibilities through internal networks, seminars, and informal meetings to foster effective communications. The issue, then, is a firm commitment to making interactions an ongoing and viable activity.

Encourage feedback. Welcome questions and comments through an ongoing interactive feedback system among various groups. If you avoid such interchanges, you create a damaging vacuum by preventing new possibilities from surfacing from those who would be inspired to act. The object is to develop dialogues that trigger the creative process and open minds to think in areas not otherwise considered, as illustrated in the 3M example.

Delegate responsibility. Morale will rise if you challenge teams and give individuals responsibility. Often they will surprise you with solutions to problems that may have eluded you. Delegating also sends a strong message of confidence that you can let go and permit the team's creativity and initiative to take over.

Benchmark activities. Set high standards by benchmarking your activities and goals against the best. Look beyond your company. Look to other industries for ideas and inspiration, and show the team what can be accomplished. Try to draw meaningful parallels and let the competitive spirit flourish.

Accept diversity. Respect differences by accepting the viewpoint that others may not see things as you do. Therefore, recognize differences of opinion, especially where there are diverse backgrounds. This is particularly important as more and more interactions take place with individuals in offshore locations.

Observe performance. The object here is to stifle problems and correct situations before they fester. This process is also the basis for further training and an agenda for group meetings.

Maintain optimism. Stay positive and avoid displaying anxiety or worry. During times of intense pressure, sidestep disclosing your concerns or fears to your personnel. Instead, present a problem as a challenge that can be turned into an opportunity through a unified team effort. As for optimism, one old sage declared: "The most important thing is to be optimistic. If we fall, let us strengthen ourselves. Heroism is in the heart."*

Now the supreme requirements of generalship are a clear perception, the harmony of his host, a profound strategy coupled with far-reaching plans, an understanding of the seasons and an ability to examine the human factors.

Sun Tzu

To summarize, if you want to manage outstanding individuals, "examine the human factors" and excite their ambition. Get them to feel the stimulation of accomplishment, and feel the dynamics of "a profound strategy coupled with far-reaching plans." Make them see a future with which they can identify. And even where there may be some unpleasant assignments, they will elect to do the job. It is within those jobs that you will find high-performing employees.

In the final analysis, success is a matter of morale—and your ability to engage heart, mind, and spirit.

Unfortunately, moral factors will not yield to academic wisdom. They cannot be classified or counted. They have to be seen or felt.

Clausewitz

---

* From the teachings of Rabbi Nachman of Breslov, 1772–1810.

8

Turn Uncertain Market Situations into Fresh Opportunities: Move to the Offensive

If courageous, a commander gains victory by seizing opportunity without hesitation.

Sun Tzu

Sun Tzu made it abundantly clear that "seizing opportunity without hesitation" is expected of a leader. From his comment, it reasonably follows that for any opportunity to materialize, the process begins with a manager's mind-set, which becomes the starting point for developing a strategy action plan. The mind-set, then, *is* the driver of actions.

For instance, think of the mind-sets of the following prominent individuals of past decades who seized opportunities under conditions as grueling as the ones that exist today.

David Sarnoff of *RCA* commanded what was called an Invisible Empire of the Air, intangible, yet solid as granite. The company, under his guidance, became the foundation for today's electronic mass media.

Early on, Sarnoff saw opportunity by marrying wireless communication to the then-emerging industry of radio. He devised a bold plan to take a fractured industry and stitch it together into a chain of stations to form the National Broadcasting Company. Thus, a program created in a New York studio could be piped simultaneously to stations in Texas, California, and Maine.

More strategically, he broadened his range of opportunities by envisioning radio and the evolving technology of television as public entertainment. When David Sarnoff died in 1971, the communications industry, which had begun as a start-up with wireless technology scarcely a century before, had revolutionized human life and made an incredible impact on how people viewed the world.

John H. Johnson, founder and owner of a complex of holdings including *Ebony* and *Jet* magazines and two cosmetics companies, exemplifies the punishing journey from fragile start-up, through mountains of racial barriers, to powerful industry leader. Johnson's success demonstrates a bold and tenacious mind-set, which he expressed as: "When I see a barrier, I cry and I curse, and then I get a ladder and climb over it." It was his positive attitude, determination, courage, and ambition that formed the psychological frame of mind needed to win the competitive onslaughts he faced throughout his life.

Long before experts spoke about opportunities through niche marketing, Johnson recognized how the power of concentration could maximize the impact of an organization's resources and successfully challenge industry leaders. He also grasped the economic potential of the black American market, which during the 1940s, 1950s, and 1960s was largely ignored by the mainstream companies.

His first magazine, *Negro Digest*, was modeled after *Reader's Digest*. His next magazine, *Ebony*, was developed in the image of the then hugely successful *Life* magazine. By 1983, it was estimated that all of Johnson's magazines reached nearly half the U.S. adult black population. He also became the first black man to land on the *Forbes* 400 list.

Mary Kay Ash, founder of *Mary Kay Cosmetics*, launched a direct-sales cosmetics company in 1963 when she was 45 years old. It was also a time when most American women did not hold full-time jobs. From a modest beginning at a small Dallas storefront and a $5,000 investment, the start-up evolved in just 13 years to become the first company chaired by a woman to be listed on the New York Stock Exchange.

Mary Kay's success is based on sound business strategies combined with tenacity and original thinking about the marketplace and about women themselves. She pieced together a corporate culture centered on education, participation, and empowerment of women. "I wasn't interested in the dollars-and-cents part of business; my interest in 1963 was in offering women opportunities that didn't exist anywhere else," she wrote later.

Her vision resulted in an entrepreneurial culture that inspired hundreds of thousands of Mary Kay saleswomen to become, in effect, small-business operators. That approach consisted of a home-spun management style whereby Mary Kay boosted their self-esteem and confidence through constant positive reinforcement and material reward. Mary Kay's "Ladder of Success" had rungs beginning with ribbons and bracelets, and reached to the highest level of the ladder, the coveted pink Cadillac.

Robert Noyce and Gordon Moore established *Intel* in 1968. They left the security of a large, established firm to start their own company. Their plan was to manufacture a product they had not yet invented: a tiny semiconductor chip with the same capacity to store computer memory as the large magnetic cores used in mainframe computers. The chip, called the microprocessor, is often ranked with McCormick's reaper and Henry Ford's assembly line as a milestone in the history of invention.

In more recent decades, other executives and entrepreneurs have seized opportunities that have made indelible marks on society, with such phenomenal successes as Microsoft, Google, Facebook, Starbucks, Dell, and numerous others.

> If a general is not courageous, he will be unable to conquer doubts or create great plans.
>
> **Sun Tzu**

What were the commonalities among those individuals? They all exhibited courage supported by a churning inner strength "to conquer doubts or create great plans" that kept them from faltering during the harrowing growth stages of their businesses. They forged a collective mind-set characterized by ambition, courage, and tenacity. The output resulted in "great plans" founded on one indisputable strategy principle that has prevailed with steadfast certainty throughout military and business history: *When on the defensive, plan for the offensive.*

Expressed another way, if you act defensively to protect a market position from an aggressive competitor, take the preliminary step of moving to the offensive. As for the alternatives, being stalled by lack of ideas, immobilized by fear and doubt, standing still, or restrained by blurred imagination can fester into severe problems. In all likelihood, your competitor feels no less anxiety when faced with similar choices: React or retreat. Thus begins a mental sparring game with choices vacillating from firm resolve to paralyzing fear.

The essential point: If entangled in a tough competitive situation, rather than languish in indecision, it is in your best interest to develop a proactive posture.

Further, as rivals see you initiate actions in an organized and consistent manner, your actions make a striking impression. They signal your determination to shift the psychological advantage in your favor.

Therefore, maintain the momentum by preplanning a variety of ready-to-use initiatives, such as launching a new or modified product, delivering an add-on service, kicking off a preemptive promotion to high-potential segments, or taking advantage of emerging, neglected, or poorly served market segments.

Otherwise, by avoiding such planning, you fumble between fight-or-flee decisions. And if you habitually back away or remain overlong on the defensive, you end up overlooking ripe opportunities when they arise. Just as serious, you send a clear message to competitors that you are not going to challenge any of their aggressive moves.

Equally severe, over the long term, your entire operation could be at undue risk. "If you don't take a risk on a new idea, that in itself becomes a risk," asserts President Tsuyoshi Kikukawa of *Olympus Corporation*.

Consequently, planning for the offensive dictates that you sharpen your ability to monitor competitors' actions in such areas as price movements, product innovations, promotion drives, or supply-chain initiatives. It also means taking the pulse of the market and observing the dynamics of your customers' buying behavior.

Doing so will give you time to organize your plans and shape your strategies. Consequently, at times you hold back, checking your own inclination to act impulsively, at other times giving it free rein.

Going on the offensive provides forward motion, which in itself takes on the spirit of determination. It is the combination of offensive action and boldness that should define your highest level of accomplishment and underscore your primary challenge; namely, develop business strategies that ultimately outthink, outmaneuver, and outperform competitors.

Further, the higher you rise on the executive ladder, the more likely you will face situations that require you to act with some measure of daring. It is at this level that you have to support your moves with precise estimates of available resources and a flow of reliable competitor intelligence. In that way you can shape your business plans that seek out opportunities and prevent pointless actions based on guesswork and emotion. By this means, boldness and courage become prudent and responsible acts of leadership.

> A distinguished commander without boldness is unthinkable. No man who is not born bold can play such a role, and therefore we consider this quality the first prerequisite of the great leader.
>
> **Clausewitz**

History teaches that an enterprise, accented with determination and purpose, leads more often than not to successful performance. Other factors being somewhat equal, when boldness meets caution, boldness wins; this quality is "the first prerequisite of the great leader."

Audacity in plans and action, therefore, has a powerful emotional impact on the mind and subsequent behavior, whereas excessive caution results in being handicapped by a loss of stability, initiative, and momentum. Thus, when in doubt, some action is better than no action. The following case provides a tangible perspective to this personality trait.

> Opportunities in war don't wait.
>
> **Pericles**

Apple Inc. had a long history of being a niche player in the personal computer market. At one point, when overall market sales reached a plateau, Apple's share hovered in the paltry three percent range, which it held for several years.

When presented with the uncertain consequences, yet driven by the need to grow or go downhill, the late Stephen Jobs saw "opportunities [that] don't wait." He made the bold leap out of his comfort zone and moved aggressively into the mainstream of the consumer-electronics market.

Apple launched the iPod. Original industry estimates pegged sales at 12 million units, with sales growth predicted at a rate of 74 percent annually for several years. What initially gave the iPod a boost was its ability to work with most PCs. As a further advantage, the product was at the introduction stage of the market and product life cycles.

In this way, by moving rapidly, Apple gained a solid market position and avoided fighting costly battles against slower-moving competitors attempting to latch on to the spiraling product category.

Then, while still in the growth stage and where market leadership for such music devices was still up for grabs, Apple sustained its momentum by establishing brand identity and thereby solidifying its competitive lead. Once on the move, Apple again made industry headlines by rankling leading cell phone makers with its innovative iPhone and later with the iPad.

> Knowing when not to fight is as vital as knowing when to accept battle.
>
> **Sun Tzu**

What can you learn from the Apple case? Even for individuals who do not have the innate personality and temperament to act with boldness, some measure of audacity still remains a practical, and even prudent, course of action. That is, as long as the display of courage is supported by a sound business plan, ongoing competitive intelligence, clearly stated objectives, and skillfully crafted strategies. Then your boldness is justified and you vastly improve your chances of extending your horizons, with the strong likelihood that you will end up with profitable outcomes.

Conversely, taking mindless risks by entering markets that are indefensible and beyond your company's financial resources, supply-chain strengths, and employees' skills are contrary to the intent of moving to the offensive. The pragmatic reality does exist, however, whereby you must face up to a market situation that is time sensitive and requires you to take some action or forfeit a lucrative opportunity to a competitor.

That means, for instance, having to live with gaps in competitive intelligence. In such a position, you have little choice but to move ahead with the information on hand, relying on your experience, self-confidence, flexibility, and intuition to deal with market conditions as they unfold.

"Sometimes you have to operate on instinct and fire before you have all the information on hand. That can flush the opportunities out of hiding," declares CEO Masamitsu Sakurai of the *Ricoh Company.*

Thus, the impulsive display of boldness is not an isolated act. Rather, it is an ordered and calculated approach that reduces the level of risk embedded in most audacious actions—so that "knowing when not to fight is as vital as knowing when to accept battle."

> The wise general in his deliberations must consider both favorable and unfavorable factors. He ponders the dangers inherent in the advantages, and the advantages inherent in the dangers.
>
> **Sun Tzu**

As part of your planning process to deploy human, physical, and financial resources to implement your plan or encounter a competitor, the following pages detail the five preliminary steps you should take. They provide you with a systematic approach to ponder the "dangers inherent in the advantages and the advantages inherent in the dangers" when making any decisive moves. These include

Make reliable estimates and calculations.
Use a diagnostic tool to assist in clear thinking and to prioritize your objectives and strategies.
Hold reserves to seize opportunities.
Assess levels of creativity and innovation.
Evaluate the ability of your personnel to deal with friction originating within the company and from the marketplace.

The general who loses a battle makes but few calculations beforehand. Thus do many calculations lead to victory, and few calculations to defeat; how much more no calculations at all! It is by attention to this point that I can foresee who is likely to win or lose.

Sun Tzu

ESTIMATES AND CALCULATIONS

Keeping in mind Sun Tzu's guidance that "many calculations lead to victory," what exactly should you estimate? Your best approach is to focus on the following five, all-encompassing categories: *internal and external relationships, leadership, seasonal forces, market selection,* and *policy.*

Look at the interpersonal relationships that exist among individuals within your business unit or product line. Determine if they are motivated to act with some measure of enthusiasm and self-assurance. Or are they sluggish, bored, and fearful of personal risk?

Are they managed by you and others with respect, openness, and honesty? Does the existing corporate culture unite them in purpose, staying power, confidence, and a sincere willingness to overcome the inevitable competitive obstacles? Is there a working environment that encourages creativity and the forward thinking that results in innovative products and services? (See Chapter 7, "Develop a Morale Advantage.")

Observe carefully the caliber of *leadership* in this estimate, which is often pivotal to the outcome of a competitive encounter. One definition cites leadership as influencing people by providing purpose, direction, and motivation while operating to accomplish the mission and improve the organization.

Chapter 6, "Master Leadership Skills," deals more extensively with a manager's positive qualities. The following is a modified list of those qualities to match the content of estimates and calculations.

> If *perceptive*, a manager recognizes from the onset the eventual impact of new competitive entries, changing customer behaviors, industry restructuring, and environmental influences. Once noted, an incisive leader acts rapidly and boldly to create opportunities and diffuse threats.
>
> If *forthright*, employees will have no doubt how and when rewards and reprimands will be handed out.
>
> If *benevolent*, the manager respects people, appreciates their industriousness and toil, and empathizes with them under adverse situations.
>
> If *bold*, the manager gains market advantage by seizing opportunities without hesitation.
>
> If *stern*, the manager is demanding and dedicated to the long-term objectives of the organization. In turn, personnel are disciplined because they are respectful of all those strong-minded attributes, yet fearful of reprimands—or worse.

As for seasonal forces, this estimate focuses on the consequences of natural climate-related conditions. It influences how you manage your business within the variables of weather and logistics, such as seasonal outcomes of winter's cold and summer's heat.

For instance, what impact does weather have on such industries as home building and road construction, or obtaining materials to meet critical schedules, on installing communications systems, or on supplying energy? What about the ancillary products and services associated with those industries—and with your industry? Further, what is the weather or seasonal impact on food supplies, fashion, entertainment, and retailing?

With a *market selection* estimate, your concern focuses on the efficient movement of your products and services throughout the supply-chain to selected niches in targeted markets. Also, your estimates of economic conditions and the intensity of competition in a territory can help immeasurably in determining the ease or difficulty of serving various segments that may be a city or a continent away.

Currently, much anxiety about conducting even the most complex business transactions over extended distances dissolve with the global use of the Internet and advances in technology. In its place, attention has shifted to defining markets by closing cultural and behavioral distances.

Hyundai Motor Company, for example, pushed into the hottest emerging markets against more established rivals. In India, it reached a strong number-two position and began aiming for the top position in the small cars market. In China, where Hyundai began selling cars in earnest only in 2003, it skyrocketed to the top spot a few years later. And in Russia, it also leaped into a leading brand position.

Hyundai's strategy was to estimate and select growth markets and outflank competitors by being first to take over smallish segments. Then, once a foothold was established, managers pushed out to additional segments, supported by modern plants with economies of scale to operate profitably.

Policy covers those fundamental guidelines that control an organization— or a business unit. Policy forms a tangible imprint of your company's ethical and operating procedures. It gives your organization consistency and a distinctive personality.

While policymaking may be outside your area of input, it does influence heavily your ability to

Select the types of strategies to grow your business

Determine the parameters by which you can innovate and compete for market advantage

Attract new talent and assign others to new levels of authority and responsibility

Secure your position in the supply-chain, with particular attention to solidifying relationships and deterring competitive threats

Deploy your financial and human resources to exploit market opportunities

Policy, therefore, holds a legitimate and powerful grasp on your business plans. It impacts your ability to cultivate the natural growth of the market. And it controls the strategy options you can use to avoid hostile price wars and other damaging activities, especially against competitors with little interest in nurturing the long-term prosperity of the marketplace.

It is sufficient to estimate the … situation correctly and to concentrate your strength. There is no more to it than this. He who lacks foresight and underestimates his enemy will surely be captured by him.

Sun Tzu

USE A DIAGNOSTIC TOOL

How you choose to prioritize your actions depends on how you assess your company's or group's strengths, weaknesses, opportunities, and threats. One familiar diagnostic tool, to "estimate the situation correctly" and determine how and where to "concentrate your strength," is the SWOT* analysis (Table 8.1). It is a widely used and time-tested approach, especially within the framework of shifting to the offensive.

When employed in a group setting, it provides a highly reliable technique for estimating your situation from internal and external vantage points. You also have options of adding complexity to the analysis by using a quantitative weighting system to grade each of the items you wish to evaluate. Or you can simplify the use of SWOT by referring to the points as a guide to an informal, freewheeling discussion.

Even with your best efforts at conducting an accurate SWOT analysis, the reality exists that your business plans can come apart when your original estimates about market conditions do not materialize.

Unforeseen situations can loom as potential threats, such as unexpected price wars, disgruntled channel members along the supply-chain, changing industry priorities, or shifting demographics in your primary segments. In addition, there are threats from overly aggressive competitors that can hamper your best efforts.

Therefore, build enough flexibility and what-if scenarios into your business plans. Also, develop second-tier objectives in the event the primary ones are no longer within reach. Then you can react in sufficient time to remedy situations that might otherwise deteriorate beyond a reasonable chance of recovery.

---

* In addition to SWOT, use the evaluation tools provided in the appendices to this book: *Strategy Diagnostic System* and *Appraising Internal and External Conditions*. If you find other tools are readily available that would serve your company's needs, use them.

TABLE 8.1

SWOT Analysis

	Strengths, Weaknesses, Opportunities, Threats
Strengths	Look objectively at your organization's or group's unique strengths, not just its physical resources.
	Identify those special skills inherent in your organization that would permit you to push the boundaries of innovation and discovery.
	Single out any unique characteristics in corporate culture, leadership, internal communications, products, systems, technologies, and processes.
	Now, do the same analysis about your competitor.
Weaknesses	Determine what weaknesses you see among the above factors.
	Look at possible choke points that could prevent implementing business plans.
	Examine what can be revamped, reorganized, or discarded.
	Estimate at what costs in time, money, and resources drawbacks can be remedied.
	What weaknesses can you detect about your competitor?
Opportunities	In scanning the customer, competitor, industry, and environmental situations, what opportunities do you see that would represent a decisive point?
	What openings exist to displace the competition, expand the company's entry into new markets, serve new customer groups, or generate new revenue sources?
	How would you define the opportunities for long-term growth vs. short-term limited payouts?
	What similar opportunities would your competitor have against you?
Threats	What immediate and longer-term threats do you anticipate, and how are you going to face them?
	Are advances in technology outpacing your company's financial and human capabilities to keep up?
	What governmental, environmental, or legislative issues are looming to hinder your ability to operate profitably?
	What point of concentration can competitors use to threaten your market position?

In any specific action, we always have the choice between the most audacious and the most careful solution. Some people think that the theory of war always advises the latter. That assumption is false. If the theory does advise anything, it is the nature of war to advise the most decisive, that is, the most audacious.

Clausewitz

Consistent with the strategy of shifting to the offensive, Clausewitz reinforces the point by advising that if given the "choice between the most audacious and the most careful solutions," take the "most decisive, that is, the most audacious," pathway to activate your plans. All things considered, then, when activating your plans, you can take an audacious course of action in such areas as enhancing your product's competitive position with new applications or launching into emerging market niches to open new revenue streams.

As for additional possibilities, you can push pricing decisions down to field personnel, initiate fresh promotional incentives, establish joint ventures to access new technologies, and install innovative value-added programs within the supply-chain.

As to what constitutes *bold,* you would have to determine that by looking at the historical patterns in your industry or your marketplace, observing the general buying patterns when new innovations are introduced, and observing the nature of the competition at all levels of the marketing mix. Again, to keep a sense of balance, keep in mind that these factors are bound to a guideline of "when on the defensive, plan for the offensive."

In any circumstance, it is an unacceptable alternative to stand still, immobilized by fear, and see your business plans falter or to take the risk of giving your competitors open access to your markets and supply-chain. Then, you may have to endure the disagreeable task of going back to management with a suitable explanation and appeal for more resources—with the strong odds of being turned down.

An additional and workable solution to permit flexibility is to retain sufficient reserves. This maximizes your ability to maintain an elastic position against known and unknown circumstances. Within that same context look to your staff for answers, for the wise executive "selects his men and they exploit the situation."

USE RESERVES TO SEIZE OPPORTUNITIES

Holding reserves permits you to react positively to unexpected opportunities, such as sudden openings in previously neglected or poorly served market segments. Additionally, you open up options to react confidently when faced with a sudden threat from a competitor.

Reserves apply to a variety of choices: Portions of your operating budget can be held in reserve to make the most of the rapid adoption of your product, for instance, by funding additional advertising, offering extended warranties, or adding customer-service personnel. In addition, there are options in packaging innovations, value-added services, additional buying incentives, pricing discounts, or new promotional formats such as streaming messages over the Internet that could be timed to coincide with customer acceptance or to upset unexpected competitive moves. Further, you can organize your personnel to form a *flying reserve*, which gives you the flexibility of shifting individuals from one location to another, depending on evolving market conditions.

Not allowing for reserves is a serious omission in your business plan. You forfeit options to immediately follow up and take advantage of budding opportunities.

Exercising those options, however, is only relevant if you know how and when to activate your reserves. That requires reliable market intelligence, with particular emphasis on understanding your competitors' strategies, assessing the qualities of their personnel, determining their strengths and weaknesses, and evaluating the overall caliber of their leadership. (See Table 8.2.)

The end purpose, therefore, for holding reserves is totally compatible with the strategic rationale of what you are in business to accomplish:

1. Satisfy the needs and wants of your customers in a manner that is superior to that of your competitor.
2. Maintain the long-term growth and viability of your market(s).

TABLE 8.2

What to Look for in a Competitor

How entrenched is the competitor in the marketplace—with special attention to the supply-chain, relationships with key customers, and the stage of customers in the buying cycle, e.g., introduction, growth, maturity, or decline?
What image does the competitor communicate to intermediaries and customers? How does it compare with yours?
How would you describe the competitor's strategies: aggressive, moderate, or minimal?
Is there a strategy pattern used to launch a new product, maintain its market position, or respond to a pricing attack?
How committed is the competitor's management to sustaining an ongoing effort in the market: heavy, moderate, or light commitment?
How interested is the competitor's senior management in the long-term development of the market?
What is the overall quality of the competitor's personnel in terms of skills, training, discipline, and morale?
How would you rank the competitor's leadership? (See criteria in Chapter 6.)

Pay attention to your enemies, for they are the first to discover your mistakes.

Antisthenes

Consequently, the piercing signal for you is *not* to commit all your resources at one time. Being able to provide an energetic response to market opportunities at a time and place of your choosing places you in a superior position to turn a doubtful situation into a decisive victory. As for competition: Reserves permit you to be ready in the event the rival is the "first to discover your mistakes."

What follows is a fundamental principle and central theme of this chapter: Move to the offensive. Never remain completely passive, even when under a disadvantage from a larger company. Simply keep moving. Stay on the offensive, no matter what the effort. Just do not lose the momentum! The following case illustrates a practical application for holding reserves.

Arrow Electronics is a global provider of products, services, and solutions to 120,000 original equipment manufacturers, contract manufacturers, and commercial customers through a global network in 53 countries. At one point, management faced the daunting problem that some customers, such as the huge contract manufacturers like *Flextronics*, bought more parts directly from suppliers and thereby cut out the middleman, such as Arrow.

After assessing the situation, Arrow recognized that it excelled in one area that represented a huge reserve, as well as a core competency, which was not available from direct suppliers: the ability to provide superior services.

For Arrow, services included generous financial arrangements, on-site inventory management, parts-tracking software, and chip programming. They also provided software that helps customers identify parts that were easily available, soon to be obsolete, or that could be made according to new environmental standards.

Consequently, the ability to provide services in and of itself is not unique. But if used to confront suppliers that usually rely on middlemen to provide a wide variety of services associated with sound customer relationship management, then it becomes a formidable counter-strategy.

In one instance, Arrow voluntarily kept three material planners at a customer's location to handle parts flow. They were also alert to look for ways to substitute parts that Arrow could supply for less. The result was that total procurement costs for some of Arrow's customers dropped by 20 percent, and new products got to market in just seven months versus the 15 months previously.

In 2012, Arrow was recognized as one of Fortune's World's Most Admired Companies, marking the twelfth consecutive year that the company has been named in the annual list.

In sum, reserves mean that you can shift to the offensive against competitors when and where needed. You thereby gain the advantage of choosing one point in your competitor's weaknesses in which to concentrate your efforts.

> Given the same amount of intelligence, timidity will do a thousand times more damage than audacity.
>
> **Clausewitz**

ASSESS LEVELS OF CREATIVITY AND INNOVATION

At the turn of the twenty-first century, companies were thought to be secure if connected to the Knowledge Economy as a competitive advantage. However, that too has become commoditized and moved, in part, to China, India, and some east European countries.

What has evolved for many companies is a shift from the Knowledge Economy to the Creativity Economy. The movement has powerful implications for sustaining the offense and developing it into a core competence.

And it is not just by maintaining an edge in technology; the movement applies creativity, imagination, and innovation to any functional areas of the business that could impact offensive strategy. Nor does the movement appear to be a short-term fad.

Increasingly, companies are embracing the change. Universities are introducing numerous programs in creativity. And a growing number of design labs are catering to companies hungering for dynamic paths for growth. The movement aims at creating new consumer experiences, not just modifying existing products or introducing line extensions.

At the forefront of the movement are *Procter & Gamble* and *General Electric*, followed by *Apple, 3M, Microsoft, Sony, Dell, IBM, Wal-Mart*, and scores of others. These companies are driving forward, armed with an intimate understanding of consumer behavior. They are honing their ability to determine what consumers want even before people can articulate these needs.

A process has evolved to foster creativity. It consists of the following steps:

Maintain ongoing observation. Go beyond the conventional market research studies and get into the streets, talk with customers, and observe their buying habits. Cultural anthropologists call the process an ethnographic study. Applied to business, for instance, managers at *Gap Incorporated* observed that social shopping—in pairs or threesomes—is the norm for women shoppers in its stores. Noting the consistency of such behavior, Gap management enlarged dressing rooms to accommodate that buying pattern.

Create models, videos, or simulations. Using a hands-on, interactive experience permits concepts to come alive. The feedback helps designers decide what to modify or discard. It thereby reduces the risk of failure and quickens the product launch.

Develop a narrative. Designers have found that wrapping a potential new product around an emotional story connects with consumers and improves the chances of success. The design of a new line of watches and driving shoes, for example, captured the story of the Mini Cooper's cool urban driving experience. The happening related to the driver, not the car.

Install a process. The object is to make the creative process an ongoing and intrinsic function of the organization. That means understanding the culture of the organization and undertaking changes where appropriate. Sometimes the changes can be wrenching experiences, but the potential payout is enormous. As business consultant Bruce Nussbaum termed it, the Creativity Economy is more anthropology and less technology.

While creativity, imagination, and innovation apply most often to new products and services, you can also activate the process and add originality to your business strategies and give them a unique character. You thereby ensure that you are elevating your thinking to a new dimension and not simply repeating the off-the-shelf actions of a past period (see Chapter 9).

> By taking into account the favorable factors, he makes his plan feasible;
> by taking into account the unfavorable, he may resolve the difficulties.
>
> **Sun Tzu**

FRICTION WITHIN THE COMPANY AND FROM THE MARKETPLACE

As you shift to the offensive and ready your plan for implementation, you must reckon with the realities of a competitive marketplace "by taking into account the favorable factors" as well as "the unfavorable." In particular, it is the unfavorable market forces that can create *friction* and the likelihood that your business plans are delayed, modified, or totally trashed.

The eminent business scholar Michael Porter[*] lists his famous five competitive forces, which are characterized here as potential types of friction:

1. Rivalry among existing firms in an industry
2. Threat of potential new entrants
3. Bargaining power of buyers
4. Threat of substitute products or services
5. Bargaining power of suppliers

Friction takes other inhibiting forms. You will find the damaging effects in the psychological sphere when exhibited by employees through

Low morale
Fear and uncertainty
Lack of trust resulting from ineffectual leadership
Depleted levels of energy due to negative perceptions about unfolding market events
Discouragement, and even defeatism, resulting from aggressive competitive actions

Friction also springs from apathy and buying resistance from customers to your product offerings, or indifference to a new incentive program from key players in the supply-chain. Then there is deep-rooted friction from inexperienced or poorly trained employees. They are the ones who are not up to the rigors of implementing offensive strategies, which require discipline, cooperation, commitment, and mental agility to stay balanced against the gyrating ups and downs of competitive encounters.

---

[*] See Porter, M., *Competitive Strategy, Techniques for Analyzing Industries and Competitors.* New York: Free Press, 1980.

Consider, too, the internal friction from the organizational logjams and layers of management that prevent the clear communication of directions from senior management to field personnel.

Also, due to an inability to obtain correct data in a usable format, friction surfaces when decision-making managers are unable to correctly estimate the situation and act rapidly on a market opportunity. Still other areas of tension come from the internal staff that fails to provide timely financial, legal, logistical and other vital information.

Friction continues its insidious damage by fostering errors that slow down day-to-day operations. With the causes of friction seemingly limitless, you should be fully aware of the ones that surround you. Then do what is necessary to limit the irreparable damage to your overall business plan as you shift to the offensive.

> The power to recognize your chance and take it is of more use than anything else.
>
> **Machiavelli**

The "power to recognize your chance" has as its foundation the conscious use of logic, experience, and training. Beyond those essentials are the immensely powerful personal qualities of the inner mind to deal with the potential damaging effects of friction: namely, *intuition*.

> To assess things in all their ramifications and diversity is plainly a colossal task. Rapid and correct appraisal of them clearly calls for the intuition of a genius.
>
> The inner light ... the inward eye ... discreet judgment ... unerring prescience ... the sensitive instinct. It is a higher form of analysis. Action can never be based on anything firmer than instinct, a sensing of the truth ... that the mind would ordinarily miss or would perceive only after long study and reflection.
>
> **Clausewitz**

To some, intuition suggests an ethereal quality that cannot be pinned down when it comes to developing actionable strategies and reducing the dire effects of unfavorable situations. Yet, there is sufficient historical and empirical evidence from the likes of Clausewitz to modern-day leaders to confidently rely on this innate quality to take action and avoid being immobilized by friction. This is especially so where rational thinking and market intelligence do not produce trustworthy solutions.

You are able to experience intuitive assistance in a variety of ways, such as by instinct, insight, hunch, or gut-feel. You also receive impressions in the form of a vision, hearing, and sensation.

Therefore, in a competitive confrontation, what ultimately plays out is a contest of your mind's creativity and originality against that of a competitor's mind. Consequently, where you need to rely on intuition, you can engage the mind and free it in a purposeful direction. Intuition, therefore, is personal and takes on your inborn personality, as your mind goes to work on a problem.

Strategy, then, is a blend of art and science that embodies the distinct imprint of the individual, which is made distinguishable by the infallible quality of intuition. Therein lies the genius of those managers who overcome friction and rise to success.

For some strong-willed, in-charge managers, nothing can replace intuition. And even where managers lack originality and a determined personality, there are decisive moments when they must take counsel within themselves, make decisions, and move forward.

Accordingly, trusting in intuition to contend with friction is a reliable technique. And seasoned managers consciously know the value of intuition in emergencies. Yet, they are also fully aware that intuition must be anchored to solid experience, judgment, and ongoing training—as well as to the concepts and principles suggested in this and previous chapters.

Proof that your intuition is working appears when you reach a comfort level and where relatively sound decisions come almost automatically—so that you know intuitively, for instance, that one strategy is more likely to work, whereas another will not.

And notwithstanding the voluminous quantities of knowledge to support decision-making, savvy managers understand that most market events are more or less hidden in a mist of uncertainty. And uncertainty is further magnified as they recognize that competing managers must rely on intuition, as well. Moreover, those rival managers are also surrounded by dynamic physical and psychological forces that create damaging friction that clouds the competitive scene.

> A skilled commander selects his men and they exploit the situation. Now the valiant can fight; the cautious defend, and the wise counsel. Thus, there is none whose talent is wasted.
>
> **Sun Tzu**

Although your weightiest decisions are often made on uncertain premises, it would be totally false to assume that success is a matter of sheer luck. It is not luck in the ordinary sense that brings achievement. In the long run, so-called chance favors the skilled and intuitive manager. This is especially so when you select people to "exploit the situation [so that] … there is none whose talent is wasted."

With that in mind, there are five propositions with which you should be familiar. If you apply them, you will increase your chances of success:

1. Seize opportunities by holding firm to a mind-set that takes a bold, proactive approach, rather than taking a more cautionary move.
2. Make reliable estimates and calculations of resources, markets, and competition as part of your strategy development.
3. Hold reserves and select your people to exploit favorable market opportunities, as well as to react quickly against aggressive competitive moves.
4. Encourage creativity and innovation in your role as a leader.
5. Recognize and deal with friction originating within the company and from the marketplace.

In sum, taking all actions that permit you to shift to the offensive is one of the most productive and winning principles of strategy. Therefore, even when forced to the defensive, as in protecting your share of market, your best course of action is to develop a plan for the offensive. As Clausewitz pronounces:

A fundamental principle is never to remain completely passive, but to attack … frontally and from the flanks, even while he is attacking us.

Action in war is like movement in a resistant element. Just as the simplest and most natural of movements, walking, cannot easily be performed in water, so in war it is difficult for normal efforts to achieve even moderate results.

Clausewitz

9

Think like Strategists: Lessons from the Masters of Strategy

All men can see these tactics whereby I conquer, but what none can see is the strategy out of which victory is evolved.

Sun Tzu

Apple unveils an iPod Shuffle half the size of its previous model, but able to play twice as many songs.

Wal-Mart sells a package to doctors for electronic medical records through its Sam's Clubs.

Charles Schwab provides a one-year, no-fee consultation service to select clients.

SAS, *Dell*, *IBM*, and *Oracle* market their data-mining expertise to help medical providers perform detective work and improve care.

What do those companies have in common? First, in all instances, they launched a new product or service during a severe recession, overlaid with intense global competition. Second, those companies made use of a powerful driving force, the human element, which became the foundation of all their energetic activities. That is, all actions that preceded and followed their product and service introductions were handled by individuals with a positive mind-set. Third, those individuals proved responsive to fresh ideas, innovative about developing business-building strategies, and enthusiastic when implementing their plans. And they succeeded in their actions in spite of harsh competitive conditions.

Finally, managers also displayed ample expertise in developing business plans that were aligned with their respective corporate cultures. They developed strategies that incorporated an indirect approach, surprise, speed, and comparative advantage—all of which were directed at decisive points in the marketplace. The combined effort was driven by effective leadership that focused on their employees' heightened morale.

During a similar period, the business press featured an overflow of brilliant case examples about the ongoing successes of such companies as *Google*, *Facebook*, and *Amazon*. Beyond the superior application of technology, those firms exhibited many of the attributes listed above.

In contrast, the scenarios for the more cautious and reticent organizations during those difficult economic times dictated approaches triggered by cost-saving plans, such as halting new product introductions, shelving costly projects, reducing marketing expenditures, and the ultimate step, laying off personnel.

Initiating survival strategies in a tough economy is one thing, and quite acceptable, where few new ideas materialize, or where managers and staff are immobilized by excessive fear, caution, and lack of skills. Nevertheless, it is no great comfort when you consider that growing a business is the true measure of a company's long-term progress. Retreating is not what you bargained for.

Also at the time, among those successes, there was also a dark side to the then Internet revolution, with numerous dot.com failures. On examination, and notwithstanding substantial funding from venture capitalists, it did not take long to see that many had incomplete, or at best, shabby business plans, ill-conceived marketing strategies, vacuous corporate cultures, lack of even rudimentary competitor and customer intelligence, and poorly defined customer segments.

> War is a matter of vital importance … the road to survival or ruin. It is mandatory that it be thoroughly studied.
>
> **Sun Tzu**

In this concluding chapter, what can we sum up from the above examples? What principles, concepts, even theoretical propositions can be singled out about running a business in today's challenging marketplace and which are "a matter of vital importance (and should) be thoroughly studied"? What can be found in the writings of the masters of strategy to help you think like a strategist? The following quotes demonstrate some unfailing truths.

> As water has no constant form there are in war no constant conditions.
>
> **Sun Tzu**

War is more than a true chameleon that slightly adapts its characteristics to the given case.

Clausewitz

The references to "no constant form" and "chameleon" are useful metaphors for the constantly changing environmental, economic, and competitive conditions of the marketplace. Yet, even accounting for the unpredictability of events, each campaign needs the structure of a plan to contain the great number of variables that exist in every situation.

Some variables can be isolated and measured; others cannot be reliably controlled. For instance, there are the sometimes amorphous factors of judgment, experience, and intuition. Beyond those are the even more subjective variables of luck, chance, friction, morale, and the paradoxical nature of risk that need to be assessed. And laced through the whole process is the even more elusive factor of human behavior at different stages of an event.

Even market intelligence can be a variable depending on its reliability, timeliness, and applicability for developing strategies. Thus, the conduct of competitive business is based less on a formally developed theory and more on intuition, experience, and an understanding of the rules or laws of action, reaction, and flexibility.

To that end, it is incumbent upon you to exploit an advantage through continuity of effort, even within the uncertainty of the "fog of war." Therefore, keep your competitor under continuing surveillance and pressure. If you maintain the initiative, it denies the opposing manager time to regain his equilibrium and renew his resistance.

That means that every pause between success and the next push gives your competitor a fresh opportunity to seek help from somewhere else and continue to oppose your efforts. Maintaining ongoing pressure, therefore, helps secure success at the least cost.

There is still another factor that can bring action to a standstill: imperfect knowledge of the situation. The only situation a commander can fully know is his own; his opponent's he can only know from unreliable intelligence. His evaluation, therefore, may be mistaken and can lead him to suppose that the initiative lies with the enemy when in fact it remains with him.

What matters therefore is to detect the culminating point with discriminative judgment.

Clausewitz

To win victory is easy; to preserve its fruits, difficult.

Sun Tzu

A well-thought-out business plan is an essential prerequisite to committing any amount of resources. (See Appendix 3, a proposed format for a Strategy Action Plan, which you can customize for your own use.)

As part of the planning process, you should be aware that every campaign should have a "culminating point." That is the point at which further action could prove costly and counterproductive.

One key conclusion is associated with this concept. Avoid entering several markets at the same time, for the following reasons. Consider the potential threat of "imperfect knowledge of the situation" when entering more than one market at a time, such as facing the resistance of a different set of competitors in each market and determining each culminating point. Then think seriously about the risk of spreading thin your human, financial, and material resources by entering multiple markets simultaneously.

A possible exception to this guideline is the conscious decision, after suitable research and estimates of competition and other relevant factors, to roll out a product without differentiating the offering, thereby treating all segments as equals with a one-size-fits-all approach.

> With many calculations, one can win; with few one cannot. How much less chance of victory has one who makes none at all! By this means I examine the situation and the outcome will be clearly apparent.
>
> Therefore I say: Know the enemy and know yourself; in a hundred battles you will never be in peril.
>
> When you are ignorant of the enemy but know yourself, your chances of winning or losing are equal.
>
> If ignorant both of your enemy and of yourself, you are certain in every battle to be in peril.

Sun Tzu

> No one starts a war—or rather, no one in his senses ought to do so—without first being clear in his mind what he intends to achieve ... and how he intends to conduct it.

Clausewitz

Effective management requires well-honed leadership skills to implement even the most basic competitive strategies. That means developing coherent objectives and strategies and "being clear in his mind what he intends to achieve." In turn, that requires a rational approach to estimating the situation and using a meticulous correlation of ends and means that focuses on a decisive point in the marketplace.

Sales, profits, return on investment, and other financials are well-established points for measurement. Others include customer satisfaction levels, customer retention rates, and corporate image. More difficult to measure are the flexibility to react with speed to an emerging market trend, degree of preparedness against aggressive competitors, levels of employee morale, and effectiveness of customer relationships. Also to be considered are measures covering operational efficiencies, effectiveness of internal communications, and progress on innovation.

Similarly, to the extent that you are able, attempt to acquire comparable information about key competitors. Then you can gain an all-around view of your business, which introduces a balanced assessment beyond the standard financial numbers. In the process, you can flag any underlying causes of an internal procedure or a market event, thereby getting an early warning signal to make rapid adjustments.

> One mark of a great soldier (or strategist) is that he fights on his own terms or fights not at all.
>
> **Sun Tzu**

> One ... question is how to influence the enemy's expenditure of effort; in other words, how to make the war more costly to him.
>
> **Clausewitz**

A manager who understands the nature of competitive actions should also be able to identify his or her comparative advantage and thereby fight "on his own terms." There are a few points worth noting. First, comparative advantage applies to using one's own strengths against an opponent's weaknesses, rather than attempting to match the rival's capabilities. Second, an equally important point in the search for a comparative advantage is to make the campaign relatively more expensive for the rival. Consequently, given the dynamic nature of competition, determining strengths and weaknesses is the primary approach to developing comparative advantage, which leads to superior strategies and tactics. (See Appendix 1 for a "Strategy Diagnostic System" and Appendix 2, "Appraising Internal and External Conditions.")

Another tool that is useful for comparative advantage is *benchmarking*, which serves a different purpose. Benchmarking allows you look to other industries and companies that excel in specific areas. It can apply to anything from production rates and defect levels to how you answer an inquiry.

First popularized in the 1970s by Xerox, benchmarking assesses your own performance, compares it with others, and if they are superior, helps you find out what it takes to match or exceed that level. This form of comparison broadens your perspective as you gain insight into how other companies operate. It thereby becomes not only a benchmark for your operation, but also stimulates fresh thinking and new ideas that can be adapted to give your operation a competitive advantage.

> In war the result is never final. Even the ultimate outcome is not always to be regarded as final. The defeated state often considers the outcome merely as a transitory evil, for which a remedy may still be found in comparative advantage considerations at some later date.
>
> **Clausewitz**

> Generally in battle, use the normal forces to engage; use the extraordinary to win.
>   Now the resources of those skilled in the use of extraordinary forces are infinite as the heavens and earth, as inexhaustible as the flow of great rivers.
>
> **Sun Tzu**

If you can realistically internalize the sage advice that the "result is never final" and "the use of extraordinary forces [comparative advantages] are infinite," then you can look forward to the incalculable possibilities of what ingenuity, imagination, creativity, and intuition can produce. It therefore becomes entirely reasonable to assume that each competitive situation would have many possible solutions.

Consequently, there is no need to rely on a single optimal solution. This is especially so if you unleash the creative capabilities of your staff through team participation.

In that context, then, you cannot assume that an organization's fate, its whole existence, hangs on the outcome of a single event, no matter how seemingly decisive. Thus, even after a failed campaign, there is always the possibility that a turn of fortune can be brought about by developing new sources of internal strength.

In addition to the above realities, successful managers sought ingenious ways to frustrate competitors' plans. And where possible, they even attempted to unhinge their rivals' alliances. Above all, they and their subordinates were adept at gathering, analyzing, and acting on business intelligence, with special attention to competitors' moves. They learned that ongoing business intelligence can spell the difference between success and failure in any competitive environment. (Many of the failed dot.coms discovered this truism, too late.)

Consequently, look at competition with a 360-degree view, along with a reflective view of your own business circumstances in a changing marketplace of erupting technologies, environmental concerns, and new customer dynamics. What you are likely to see is a panoramic scene that can impact your decisions about selecting markets, launching new products and services, and devising winning strategies.

> The requisite for a man's success as a leader is that he be perfectly brave. When a general is animated by a truly martial spirit and can communicate it to his soldiers, he may commit faults, but he will gain victories and secure deserved laurels.
>
> **Jomini**

> By command I mean the general's qualities of wisdom, sincerity, humanity, courage, and strictness.
>
> **Sun Tzu**

THE HUMAN FACTOR

The most significant outcomes of those winning company performances referred to earlier were the ways in which those executives harnessed their human resources. Specifically, the results related to the employees' states of mind, which reflected in their attitudes and overall ways of thinking.

The mind can be compared to a deep pool of possibilities set in motion by thoughts that come to the surface as intuitive insights. These can be positive or negative, depending on the input of the controlling beliefs. And they can be influenced to the extent to which executive leadership displays the "qualities of wisdom, sincerity, humanity, courage, and strictness."

In turn, the outcomes shape an individual's core behavior. On the positive side, insights can lead to innovative thinking that translates to competitive advantage and bold new opportunities, as in the above company examples. On the negative side, insights that are chained to fear produce pessimistic attitudes that can strangle creativity, shut down ideas, and blind the individual to problem-solving possibilities.

To assist you in taking a broad look at behavior and the possibilities of changing mind-sets and attitudes, consider the guidelines that follow. The aim, then, is to inspire those who are generally positive and compliant, the not-so-compliant, individuals with negative feelings, and even persons with insubordinate behavior.

From a managerial viewpoint, the purpose of the following review is to bind your employees' hearts and minds to "gain victories and secure deserved laurels" when they face the inevitable competitive battles that lie ahead.

Expand Your Employees' Awareness

Your initial approach is to bring your employees to a point of living in tune with their surroundings. Specifically, get them to reach an expanded state of awareness. Awareness, in this framework, takes into account the following issues: how competitors' actions could disrupt your plans, how to interpret market events and their impact on your strategies, and how to draw correct conclusions from market intelligence to create new opportunities. Therefore, your purpose is to get your employees to look outside their four walls and understand the wider competitive world. Through online communications, face-to-face informal meetings, or weekly or monthly briefings, update your staff about markets, industries, and competitive conditions. These serve several purposes, among them:

> You recognize the value of providing your staff with substantive market and competitor information.
>
> You motivate them by providing a venue so they can participate with fresh insights.
>
> You tap their diversity, thought patterns, experiences, and collective knowledge, thereby integrating individuals and their functions.
>
> You obtain viewpoints that can provide useful perspectives and constructive comparisons.
>
> You bring unity to your group.

Such sessions will be worth your time and effort. The major outcome will be that you awaken your staff to internalize that you and they walk together on a very narrow path, which includes numerous visible and unseen obstacles. The most important issue is not to be afraid. Then, if there is a balance of a positive mind-set linked with an objective interpretation of market events, success is weighted heavily in your favor.

Increasingly, a wide variety of communications vehicles permit you to take the high road and initiate programs and exchanges of information. Such is the case with the mushrooming use of social networks, blogs, e-mail, instant messaging, video conferencing, corporate wikis, and the like. Also, face time, thought to be outmoded by electronic communications, is still prevalent and noticeably on the increase.

And where there is a heightened diversity in the workforce, the need to unite individuals around a common goal intensifies. Individuals tend to be wherever their thoughts are. You want, therefore, to make certain their thoughts, as well as your own, from a business viewpoint, are where you would like them to be.

Therefore, stay close to your employees, so that you can sensitize yourself to their thoughts, moods, temperaments, and increasingly the nuances of their local cultures. All the while, your aim is to integrate them into the mainstream of your company's goals.

Doing so also allows you to focus individuals' attention on constructive projects that can take even a sketchy idea and convert it into a new product or service. Consequently, the exciting possibility exists for an embryonic innovation to sprout into a new revenue stream.

Otherwise, without some guidance, their creative thoughts would radiate like sunbeams. At first, they seem solid until you try to grasp them. You thereby lose the potential value of huge business-building resources.

Thus, reach out to the special needs of your employees and focus on sustaining their psychological well-being. As a result, you should see tangible improvements in performance, innovation, and employee harmony.

Recognize That Employees Harbor Ingrained Habits

There is substantial harm to employee morale if negative habits are left unattended. Make every effort to break those habits, which often are inclined toward damaging feelings, fears, and a lack of confidence. Do all you can to convince them not to make the same mistake as others who prematurely give up trying to change because they feel stuck in their behavior.

Demonstrate to them that if they truly want to, and are willing to work hard enough, they can overcome undesirable habits. Here is where

TABLE 9.1

Areas in which Behavioral Specialists Provide Assistance

Conduct individual and group coaching
Analyze and advise on complex personnel issues related to inner feelings and external behavior
Conduct needs assessments to define training objectives
Match the right person to the right job
Communicate company vision, objectives, and cultural values
Monitor the workforce during the development of business plans and competitive strategies
Perform damage control due to employee performance rising out of difficult competitive conditions

discipline plays a positive role. In effect, it is a test designed for them to exercise freedom of choice. The issue, then, is for them to choose wisely.

Your best approach is to set up (or recommend) a program to deal with negativity. Where possible, utilize outside specialists. Specify a clear-cut objective to help individuals resist pessimistic thoughts. Often, they can direct the process to clear the blockages that prevent implementing your business plan (see Table 9.1).

By taking such positive action to harness your human capital, you acknowledge that your people are a major influence in market performance. And, as important, they function as key competitive differentiators.

Consequently, where the profit motive normally drives major decisions, the more urgent considerations for you, and the intentions of this book, are the behavioral issues of your personnel.

> The power to recognize your chance and take it is of more use than anything else.
>
> **Machiavelli**

Learn to Wait

As noted above, maintain ongoing dialogue with your staff, so that you become aware of the direction of their thinking and mind-set. As important, use the interchange so that your staff understands your "need to recognize your chance and take it." In particular, such awareness on both sides is significant should you have to wait before implementing your plans due to sudden market events.

In that case, you are likely to find a heightened level of anxiety not only within yourself, but with your staff. This is where you, and they, must cool down and learn to wait for better timing. This is the moment, despite all determined efforts, to exercise disciplined patience.

In essence, given a choice between acceptance and anxiety, choose acceptance, with the proviso that patience is only a pause, nothing more, that allows you time to reignite your efforts with fresh ideas and innovative strategies.

Learn to Restrain Irritation

Help your people avoid being provoked by negative surroundings. That means creating compatible relationships with others. Does this imply that you have to assume the role of an active mediator? For the most part, the answer is a resounding *yes*.

Yet some managers will back off and ask the opposing groups to work through their problems. Notwithstanding, it is your obligation that once anger, frustration, and continuing dispute move forward beyond a reasonable time, you have to take an active role in setting a pathway for conflict resolution.

Still other executives intentionally create opposition in their ranks, thinking that it leads to a healthy work environment. However, it is with a degree of risk, when compared to the far greater benefit of creating harmony and unity of effort.

Nonetheless, strained relationships do exist among various functional areas of an organization, such as marketing versus finance, product development versus distribution, manufacturing versus sales, and so on.

For the most part, however, hostilities are counterproductive, especially when flaring anger dominates the scene. Clashing groups seldom arrive at an acceptable solution. Within such a fractured situation, your aim is to promote workable outcomes based on internal operating conditions, the dynamics of the marketplace, and the activities of competitors.

Then, you are more likely to come to a peaceful and healing condition among all parties—even where groups live as opposites, but not opponents. It takes a measure of sensitive awareness on your part. For it is easy to criticize others and make them feel unwanted. Any manager can do that. What takes effort and leadership skill is raising them up and making them feel good.

Look for the Good in Yourself and Your Employees

Avoid the unnecessary search for shortcomings and weak points in yourself and employees, unless there is a deliberate effort to conduct a needs analysis for future training. Otherwise, focus on their good attributes. Highlight them and turn even indifferent employees into winners. Keep in mind, too, that they are the ones who will carry out your plan, which has at stake your personal reputation—and perhaps your job.

In the process, be alert for the inevitable expressions of hardship, complaints of tough work, and sacrifices that often surface among workers during any emergency. Distinguish between what you would consider grueling work and merely a display of whining and grumbling. To counter any emotionally charged outbursts, indicate to the rest of the staff the consequences if they continue with their flare-ups.

If you succeed in choking off their complaints, and as a result of their sacrifices and hard work you show first-rate results, point out that those positive results are only the beginnings of other good outcomes.

It is your opportunity to say, "If you call this good, I'll really show you what good is. I'll show you how this company can grow and how each of you can benefit financially and professionally." Of course, this approach must be anchored to tangible and realistic assessments that the staff can understand, believe in, and rally around.

What ties into that approach are the opportunities that emerge each day. Some are quite apparent, others require more searching. Also, what might appear as sameness could be an opportunity under a different set of market conditions.

As long as there is a mind-set tuned to fresh opportunities, it is your job to instill in your people the ability to seek out possibilities of what every day and each event has to offer. You must be sure, however, that your employees see the opportunities, agree with them, and are of the same opinion.

Workers will be happy to change their individual behavior if they understand why and how their actions contribute to the overall company's fortunes. Also, they will act positively if they believe it is personally worthwhile for them to play an active role in the organization.

As part of change, utilize a workable system of rewards and recognition. Such a system includes much of what is generally accepted about positive reinforcement and classic motivational behavior. Look again at the concepts developed by Herzberg, McGregor, Maslow, and Ouchi, described in Chapter 7.

Further, as part of the process of nurturing your staff to heightened levels of performance, initiate (or recommend) procedures to capture the insights, knowledge, and observations of numerous individuals and categorize them into usable databases. Such information should be then available and easily accessed for training, mentoring, written documentation, or oral exchanges.*

Even where you cannot effect change in the organization, it is possible to take the initiative and set up a system within a small group. (See the discussion below on "Managing Knowledge.")

These points are illustrated in the following case example.

> The most complete and happy victory is this: to compel one's enemy to give up his purpose while suffering no harm oneself.
>
> **Belisarius**

Nucor Corporation is the largest producer of steel in the United States and the world's foremost steel recycler. And it does so efficiently and profitability when benchmarked against almost all other companies in the Standard & Poor's 500 stock index. Nucor's 371 percent return to shareholders over the past five years beats most other companies, including such icons as Amazon.com, Starbucks, and eBay.

What is behind such exemplary performance in an industry as Rust Belt as they come? The answers fall into several categories:

1. Performance: On average, in a nonunion environment, two-thirds of Nucor steelworkers' pay is based on a pay-for-performance product bonus, which results in the highest wages in the industry. Behind such a system is management's understanding of human behavior, which is anchored to respect, empowerment, and rich rewards for all its workers. As such, the Charlotte, North Carolina, company has never laid off a worker for lack of work. At the executive level, compensation is also tied to product performance. As an add-on, executives are measured by tangible actions aimed at beating the competition and outpacing a sample group of other high-performing companies.

2. Corporate culture: To assure a cultural compatibility throughout the organization, a managerial priority is to instill Nucor's unique culture in all of the 13 plants it acquired over a five-year period. Teams of highly experienced veteran workers visit with their counterparts

---

* In no way should this process be confused with a typical transition memo, which tends to be more limited in scope and deals with fundamental administrative procedures and the like.

in newly acquired facilities to explain the system face to face. Part of that cultural process includes explaining the deep-rooted custom of providing a helping hand without obtaining approvals from supervisors. It is a common everyday occurrence, for instance, for employees in one plant to take the initiative and help others in sister plants to get operations up and running should severe problems shut down production, regardless of time, distance, or any inconvenience. Even the cultural symbols are considered important. Every year, for example, all employees' names go on the cover of Nucor's annual report.

3. Entrepreneurship: Nucor's flattened hierarchy and emphasis on pushing power to the front line leads employees to adopt the mind-set of owner-operators. That focus gives free reign to employees to exercise their imagination and allow their intuition to flourish. For example, such commitment led to the development of thin slab casting of sheet metal that has made Nucor an industry leader. In another instance, employees in one plant had to innovate themselves out of a predicament. Its particular form of steel could not be produced profitably any longer. Relying on their personal knowledge, experience, and intuitive creativity, employees found types of specialized steel they could produce more profitably—and which would be less threatened by imports.

4. Motivation: There is unwavering attention heaped on employees. That means talking to them, listening to them, taking a risk on their ideas, and accepting an occasional failure. It is about nurturing their experiences and sharing them with others in a proactive way.

   Nucor systematically sends new workers to existing plants to hunt for improvements. Older workers also travel to newly acquired plants to find out what they can learn. These experienced individuals actively tune in to sharing ideas and experiences, as well as staying alert for innovations they can take back to their home plants. There is a healthy competition, too, among facilities. For instance, plant managers routinely set up contests to try to outdo other plants in areas such as safety, efficiency, product quality, or output. It all ties in with the company's long history of cooperation and idea-sharing.

The Nucor case provides a learning platform for you to strengthen your personal decision-making and other managerial capabilities.

The following two categories hone in more precisely on developing that capability: *managing knowledge* and *activating intuition.*

> If the mind is to emerge unscathed from this relentless struggle with the unforeseen, two qualities are indispensable: first, an intellect that, even in the darkest hours, retains some glimmerings of the inner light which leads to truth; and second, the courage to follow this first light wherever it may lead.
>
> **Clausewitz**

MANAGING KNOWLEDGE

As a backdrop to fortifying your experience, intuition, and training, you can view knowledge management (KM) as an organized and more elevated form of information. So that when you interpret a business situation, organized knowledge provides new meaning, value, and "some glimmerings of the inner light" to your decision making.

KM consists of two parts:

1. *Explicit knowledge* exists in your internal databases, records, manuals, documents, the raw numbers in spreadsheets, and increasingly through websites. In particular, sophisticated analytics incorporate additional data from a variety of customer touch points, such as call centers, field services, prior marketing programs, and back-office information. When organized into a usable format, there is greater reliability and accuracy for estimating a competitive situation.

2. *Tacit knowledge* generally resides in the minds of individuals who have accumulated it through discovery, experience, intuition, or through numerous interactions with others. Because tacit knowledge tends to be less structured, it cannot always be put down on paper. Instead, it is transferred indirectly through conversation, observation, or other types of informal interchange. Tacit knowledge can originate in a variety of patterns, such as the impressions, feelings, and insights of a sales rep returning from a visit with a key customer. Or it can start with an engineer making an off-hand comment about a gestating idea with an associate in a casual setting over lunch.

If given the same level of seriousness and discipline as any other business system, KM can operate as a balanced, multidisciplinary framework for capturing, sharing, and spewing forth immensely valuable knowledge.

Consequently, in your managerial role of making a diverse range of decisions, actively involve yourself in developing an internal KM exchange network. Regardless of the size of your company, it should blend with the everyday life of your organization and feed the transfer of meaningful knowledge, including the use of case histories and concepts discussed in this book.

For many organizations, explicit knowledge is tangible and available on a widespread basis, or minimally it is accessible to several layers of personnel. On the other hand, tacit knowledge is somewhat unbounded and tends to be used by individuals who need to protect what they know as a personal defense or a power barrier. As indicated earlier, it is this form of knowledge that is often lost to others, if not captured, organized, and made accessible.

To break through the barriers and add KM to the culture of your firm means establishing a level of trust up and down the organization and instilling a spirit of teamwork to make knowledge management work to the full benefit of the organization.

By blending explicit and tacit knowledge, more accurate decisions can evolve. For instance, you would be in a far more advantageous position to justify (or recommend to a management committee) the expenditure for developing and rolling out a new product, or for adopting a cutting-edge technology, or probing an evolving market segment. And, perhaps most importantly, and in the context of this book, to shape strategies and tactics that sustain a competitive edge.

The responsibility for KM can reside with any number of individuals, depending on the size of your company and the level of priority given to the project. The range of individuals includes company librarian, information technology (IT) manager, market research manager, or the more recent title of internal infomediary, who creates or manages systems to connect employees with the knowledge they need. (See the Johnson & Johnson case in Chapter 7.)

Regardless of the title, the central responsibility is to tap into the immense fund of knowledge flowing around the organization.

Finally, information, intelligence, and total knowledge management are only as valuable as your willingness to apply the mass of knowledge to think like a strategist, and thereby make the appropriate decisions that create a winning business advantage.

> Undoubtedly, this power of judgment consists to a greater degree or lesser degree in the intuitive comparison of all the factors and attendant circumstances; what is remote and secondary is at once dismissed while the most pressing and important points are identified with greater speed than could be done by strictly logical deduction.
>
> **Clausewitz**

TURN ON INTUITION

Much has been said in Chapter 8 about the value of intuition when faced with a dilemma and clear-cut decisions are not forthcoming. This section deals with techniques to activate this quality from within. Consequently, when pressured to make a decision with less information than you would like, which is usually quite often, you are likely to rely on "the intuitive comparison of all the factors."

Here is how to activate intuition: Think of intuition as an inner voice that presents you with possible courses of action. It unlocks the mind and guides you by means of a flash of insight, ideas, images, metaphors, or symbols. "The intuitive mind tells the rational thinking mind where to look next," declared the renowned Dr. Jonas Salk.

While intuition is often thought of as elusive, spontaneous, and outside our control, nevertheless it is possible to make intuition more accessible and more reliable. It begins by paying attention to your feelings. That means starting with a quiet mind and turning off the constant monologue that clutters the mind.

You then adopt a mind-set where simply *knowing* transcends reason. Just perceiving possibilities is also an intuitive function. Of course, you can never be certain what the outcome of a decision is going to be. But you can have a strong intuitive sense of the direction you want to pursue.

Yet such a leap of understanding is not in opposition to or a substitute for reason; and it certainly is not in conflict with the input of reliable market intelligence and a system of knowledge management.

Intuition is inner power that you use in addition to reason and factual information. In effect, there is an integration of intuition with the logical and linear thinking mind.

To make intuition work, you have to learn to hold attention on something for more than a few seconds, which can be a real challenge internally. For some, that also means emptying the mind of the emotional baggage that upsets, angers, and creates a chaotic state of mind. It means letting go and being nonjudgmental.

You may have to personalize how you reach the receptive state where intuition can flourish. It may be through meditation, solitude, being out in nature alone and quiet, or using slow breathing techniques.

The object is to quiet the feelings, quiet the body, quiet the mind, and be left with a kind of inner knowing. In the silence you learn the most about intuition. It is in this period of incubation that you let a problem just sit and you take time out.

Other approaches include using mental imagery, which is more associated with thinking and tends to be the type of intuition used by business executives, especially entrepreneurs, who tend to be highly intuitive.

In all methods, a common requirement is trusting yourself. Then you can trust your intuition. Keeping track of the accuracy of your intuitive decisions will give you an indication of whether your mind is prepped for intuition to thrive.

The benefits come when you can perceive possibilities in the future. In its most practical business application, intuition applies to developing a strategic business plan where the first planning step is to develop a mission statement for your company or business unit, or to define a long-term strategic direction for a product line.

Anything that is creative, breaks new ground, provides a future-directed vision, or pushes you beyond the boundaries of what you already know is intuitive.

If any mystery exists about intuition, it is that people seem to get information and they do not know how they got it. Mathematicians, for example, can arrive at theorems that have never been proven before, just through their intuition.

"No problem is solved by the same consciousness that created it," stated Albert Einstein who was well known for his use of intuition.

What is required is a shift in consciousness, and through this shift you tap the subconscious mind to show you the way to solve a problem, which is revealed as intuition. You then use logic, reason, and market research to follow up on the intuition for proof and validation. However, if time does not permit that luxury, then you must react and trust in your intuition.

The creative leap is always an intuitive one that enables you to see things that you have not observed before. It is a new perception, as though intuitively you notice what you have not noticed before; or you acknowledge what you already know but have forgotten. It is a sense of inner awareness rather than something that you need to learn about. Again, being intuitive is trusting yourself.

We are all equipped with intuition. It is built in. It is pragmatic. It solves problems. It identifies opportunities that may not be seen with conscious vision.

The following case illustrates the above concepts.

Whirlpool Corporation, a global leader in the home appliance industry, at one point faced a series of severe problems, notwithstanding its lofty position as the world's number-one maker of big-ticket appliances.

The central problems were that Whirlpool's machines had been reduced to commodities, and prices for its most important products were falling each year. As one Whirlpool executive explained: "I go into an appliance store. I stand 40 feet away from a line of washers, and I can't pick out ours. They all look alike. They all have decent quality. They all have the same price point. It's a sea of white."

But that explanation described only a surface issue. Underlying the situation was Whirlpool's history of giving relatively minor attention to new product development and innovation.

Whirlpool was good at operating plants and distribution channels efficiently. And it was outstanding at turning out quality washers and dryers that were solid and long-lasting. The company also achieved unmatched economies of scale with its perpetual cost-cutting.

Only on an as-needed basis did engineering tweak its products to lower costs or boost performance, such as by better insulating a freezer or adding another washing cycle. But that was about as far-reaching as product development got.

Challenged by the flagging profits and flat sales, Whirlpool made a series of turnaround moves. Its strategies focused in two areas: First, develop strategies that would prevent competitive imitators from moving its product line into a commodity status where price wars become the central marketing focus. Second, reinvent the corporate culture so that innovation and creativity become the essential strategies, as well as the triggers to energize the company plagued by missed opportunities, sluggish sales, and dismal profits.

But how is the changeover accomplished? How do you teach people to be creative? Whirlpool moved forward with the following actions:

An intranet site offered personnel a do-it-yourself course in innovation and listed every project in the pipeline. Employees were then invited to post ideas or to network informally with others to gain their expertise.

The company hosted innovation fairs to salute inventors and elicit more ideas. Senior executives continued to encourage workers to go to their bosses with proposals. Whirlpool's knowledge management site recorded up to 300,000 hits per month.

Guidelines were provided about how to present ideas that would improve the chances of surviving an executive review for funding. New ideas would have to enhance the company's existing brands or products. Only then would approved projects go to representatives from design, market research, research and development, and manufacturing departments for implementation.

Projects would be analyzed and validated through market research to determine if the new product would likely command above-average markup.

Products that failed the first go-round are viewed again for revival. The intent is not to kill ideas. Instead, these projects are shelved for a period of time so that other employees can take a look at them later in a more insightful mode of thinking.

Some of the products that have emerged include an innovative, feature-loaded waffle iron that is three times costlier than the old waffle iron, an all-in-one gas grill/refrigerator/oven/stereo/bar that fits onto the back of an SUV for tailgate parties, and a dual washer and dryer set with a variety of innovative accessories.

In five years, new products that fit Whirlpool's definition of innovation have skyrocketed. In 2012, Whirlpool won the Silver Edison Best New Product Award.

Beyond the success in achieving a resounding turnaround in its fortunes, the grandest form of compliments came from such companies as *Hewlett-Packard*, *Nokia*, and *Procter & Gamble*. These organizations, respected leaders in their own fields, eagerly sent groups to benchmark their own innovation programs against Whirlpool's.

> All ... laws and theories which are in the nature of principles are the experience of the past ... We should seriously study these lessons ... We should put these conclusions to the test of our own experience, assimilating what is useful, rejecting what is useless, and adding what is specifically our own.
>
> **Mao Tse-tung**

What can be learned from the Whirlpool case?

First, the merging of knowledge management, experience, training, corporate culture, and intuition drives creativity and innovation.

Second, as market conditions change, technologies advance, and competition hardens, many of the successful strategies, including those from time-tested historical concepts from other times and places, should be studiously reviewed for their potential value; then "put these conclusions to the test of our own experience."

Third, the challenge is for you to discover what usable information you can take away that would expand your knowledge and ability to analyze your company's situation.

Fourth, recognize that the road you must travel to carry out your business plan can never be determined beforehand with absolute accuracy. Therefore, you have to be actively engaged, visible on the scene to observe events with your own eyes, and be able to assimilate all the information. And what you see must be accurately interpreted using your innate powers of knowledge, experience, training, and intuition. Then, you can follow through with an action plan.

Fifth, conduct a series of one-on-one interviews with individuals directly connected to a campaign. Also hold group debriefing sessions to acquire as much information as possible. This effort is likely to show in striking, irrefutable facts what was actually taking place at various times and places during a campaign. (See Table 9.2.)

Going through these five steps serve as a foundation for what could lead to new innovation and strategies. In addition, that review should yield valuable insights related to changes in training or new methods of performing in the marketplace under a variety of conditions. Here is

TABLE 9.2

How to Combine Intuition, Experience, Training, and Knowledge into a Winning Force

Examine your past campaigns, events, and strategies, including those of competitors. Do not overlook the smallest detail that could help you figure out what went right or wrong in a campaign. Pinpoint those lessons that would be the basis for shaping future strategies.

Tap the experiences of individuals who can recapture events of long ago, especially those who have been elevated in the organization, or if possible from those who left the company.

Tune in to actual business case examples, past and present, in and out of your industry, for ways to enhance your skill and broaden your range of knowledge.

Anticipate that your experience, knowledge, training, and intuition will appear as an idea, image, or a flash of insight. This can occur even under unfortunate circumstances when you may have given up hope of their influence.

where a reliable business plan is essential to capture and organize all the diverse information.

> Theory cannot equip the mind with formulas for solving problems, nor can it mark the narrow path on which the sole solution is supposed to lie by planting a hedge of principles on either side. But it can give the mind insight into the great mass of phenomena and of their relationships then leave it free to rise into the higher realms of action.
>
> **Clausewitz**

CONCLUSION

If your employees get entangled in a tough competitive encounter, no doubt many will face stumbling blocks, which they will promptly pronounce as overwhelming. They may even complain that the possible solutions are too involved or overly fatiguing.

Such reactions are understandable: For some, there is the natural timidity of humans to see only one side of anything and thereby make a first impression, which usually inclines toward fear and exaggerated caution.

If, as a manager, you give in to those complaints and frailties, you will soon succumb completely. And instead of acting with courage and determination, your efforts will be reduced to weakness and inactivity.

To resist all this negativity, keep faith in your inner self. Focus on all your deep-seated knowledge, training, experience, and, yes, intuition. At times, even with all the uncertainties of the marketplace, this mind-set may appear as stubbornness. Actually, it is an expression of strength of mind and character, called firmness. Your aim is to "give the mind insight into the great mass of phenomena and of their relationships, then leave it free to rise into the higher realms of action."

The reality is that there will be the inevitable miscalculations and hurdles that you are certain to face. Notwithstanding, if you pursue your aims with boldness and confidence, you will reach your goals in spite of obstacles.

Pursue one great decisive aim with force and determination.

Clausewitz

Where the law of probability is often the only valid guide, you must stand fast; trust in your honest interpretation about what is realistically possible and what is not. Then, with some level of confidence, "pursue one great decisive aim ... with determination." This is a managerial role that is not easy to play, yet one that is mandatory for successful performance.

Equally important is the trust you must have in your subordinates. Therefore, choose individuals on whom you can rely. Such reliance and confidence is directly proportional to the training they receive and the manner in which they are led. The following example illustrates these points.

Carlos Ghosn has the mind-boggling job of serving as chief executive of both *Nissan* of Japan and *Renault* of France. He prods his employees at all levels by creating an intense feeling of urgency. If complaints crop up, he dismisses them and any other perceived difficulties with an irritated wave-off.

The urgency is intentionally created as part of Ghosn's style of anticipating problems, putting them on the table, and dealing with them before they happen. His study of corporate history shows that if you wait too long to tackle problems, you are likely to face a tragedy. "Ghosn is absolutely tenacious in fighting complacency and the notion that we are in good shape," declares Nissan executive Steven Wilhite (www.forbes.com/forbes/2006/0522/104.html).

Thus, Ghosn functions as if collapse lurks around the next corner. He is fueled by a sense of crisis, mixed with impatience and passion, as he runs, simultaneously, two of the major car companies in the world.

Operating in that cultural environment where success breeds complacency, and sometimes arrogance, is part of Ghosn's approach to elevating the creativity and productivity of workers—and thereby managing the human element.

Ghosn's leadership style provides a resounding justification for strengthening your resolve against the weakening impressions of the moment. Even fortified with personal decisiveness and armed with sound intelligence, you can still succumb to wrong decisions dictated by fear.

You must not allow these errors to shake your faith or tempt you to accommodate to those fleeting impressions. The inevitable difficulties, therefore, demand your confidence, firmness, and conviction, whereas others, "ordinary" managers, may find in obstacles ample excuses to give in.

Do not give ground in competitive encounters until the very last moment and where all options have been considered. If your plan is based on using sound strategic principles, such as boldness, an indirect approach, concentration, and speed; if you are resolute and persistent in implementing your plans, and determined to accomplish your goals, then you will find success in your efforts.

The steadfast lessons of this rule are

First, count on strengthening your decision-making capabilities by fortifying intuition, enhancing your business experience, and expanding your knowledge.

Second, look for the infallible connection of corporate culture and competitive strategy. It is the determining factor if you are to be successful in implementing your plan, directing your competitive strategy, and ultimately outperforming your rivals.

Finally, it is in the marketplace that an organization justifies its existence. Such justification, however, must be based on honesty and integrity to the foundation principles, value systems, and overall culture of your organization. It must stand firm to the long-term development of the marketplace, and not be rigged to any personal agenda or avoidance of factual market intelligence.

In the end, it is mastering the concepts and principles of strategy that can direct marketplace events in your favor.

In sum, deliberately and systematically following the enduring rules of competitive strategy can help you overcome the obstacles that have crushed other managers. Keep in mind, too, that these applications have endured over long periods in many fields of endeavor. Consequently, the more prudent path for you is to firmly understand the concepts, principles, and guidelines in this book before making a conscious decision to alter or minimize their applications.

Deliberately avoiding them altogether places you at a distinct competitive disadvantage, which could result in reducing your position to a marginal segment of the market, whereas integrating them into your business plans and strategies can increase your chances of triumphing over rivals that are looking to oust you from the marketplace.

Yet, the reality exists that developing a business strategy edge is full of complexities, such as the rational and nonrational dimensions of running a business, issues of friction within the organization that create uncertainty, and the ambiguities of luck and chance.

Therefore, some rules make sense some of the time, but all rules cannot be followed all of the time. The secret to success, then, is knowing when to break the rules and when to follow them.

> Now there are five circumstances in which victory may be predicted:
> He who knows when he can fight and when he cannot will be victorious.
> He who understands how to use both large and small forces will be victorious.
> He whose ranks are united in purpose will be victorious.
> He who is prudent and lies in wait for an enemy who is not, will be victorious.
> He whose generals are able and not interfered with by the sovereign will be victorious.
>
> **Sun Tzu**

Now turn to Appendices 1, 2, and 3 for the tools of strategy to assist you in outthinking, outmaneuvering, and outperforming your competitors.

Appendices

STRATEGY TOOLS

Know the enemy and know yourself; in a hundred battles you will never be in peril.

When you are ignorant of the enemy but know yourself, your chances of winning or losing are equal.

If ignorant of both your enemy and yourself, you are certain in every battle to be in peril.

Sun Tzu

Appendix 1. Strategy Diagnostic Tool

This tool helps assess your firm's competitive strategies against those of your competitor. It provides a reliable performance measure to "know the enemy [competitor] and know yourself." It helps you gauge the likelihood that "you will never be in peril."

Appendix 2. Appraising Internal and External Conditions

This checklist assists in analyzing key factors, such as markets, customers, and competitors, that could affect your ability to carry out your business plans. It also helps you assess your company's internal competencies.

Appendix 3. Strategy Action Plan

This tool provides forms and guidelines for you to develop a business plan based on a proven format. You can personalize the plan by inserting specific terminology and issues related to your firm.

Appendix 1: Strategy Diagnostic Tool

STRATEGY DIAGNOSTIC TOOL

Developing and implementing competitive strategies are essential to the growth of an organization. And in some instances they become critical components of a firm's survival. However, you are likely to find that challenges and obstacles do exist to hamper the strategy development process.

Some come from inside your organization—personnel, financial, or physical barriers that threaten to hold you back from moving ahead on your plans. Others result from the numerous forces coming at you in a dynamic and overly competitive marketplace.

This *Strategy Diagnostic Tool* exposes both the challenges and obstacles, and provides you with corrective actions and solutions for success. In turn, you will find yourself with an improved strategy plan to seek out business-building opportunities, as you reduce the risk natural in most competitive encounters.

How to Use the Strategy Diagnostic Tool

This diagnostic tool is organized into nine strategy guidelines. Key ideas are presented so that you can assess your organization's tendency toward effectiveness (or ineffectiveness) within each strategy. And remedial solutions are offered for each. This process provides a consistent way for you to gauge areas of needed improvement that will lead to success in your overall strategy.

Additionally, where you wish to implement changes and need detailed information, references to chapters in the text are indicated for each of the guidelines.

To get the most out of this tool and most thoroughly absorb the lessons and benefits of these nine guidelines, it is advisable to try one at a time, giving yourself and your organization a chance to roll it out, practice, and assess results.

Thus, each strategy guideline consists of three parts:

Part 1: Explanation of the guideline and items that describe if the principle is functioning *effectively* in your organization and contributing to the successful implementation of a competitive strategy.

Part 2: Items that signal if the guideline is functioning *ineffectively* and would prevent fully implementing a strategy. Used as a hands-on guide, it ensures that your organization does not lose momentum and get caught up in resource-draining choke points when it comes to successful strategizing.

Part 3: Remedial actions based on the deficiencies you uncovered in Parts 1 and 2. Included are "Executive Comments," which are real quotes from executives that tie the strategy principle to real-world problems.

To begin, ideally, using a team approach, consider how your organization rates for each item in Parts 1 and 2, and mark each accordingly as *frequently, occasionally,* or *rarely.*

Strategy Guideline 1: Shift to the Offensive

Standing still, stalled by lack of ideas and immobilized by fear, can fester into severe problems. If entangled in a tough competitive situation, rather than languish in indecision, it is in your best interest to develop a proactive approach and search for fresh market opportunities.

Part 1: Indications That Strategy Guideline 1 Functions Effectively

1. You and your staff display a proactive, shift-to-the-offensive mindset when confronted with aggressive competitors.
   Frequently_____ occasionally_____ rarely_____
2. You encourage risk taking among your staff, with no serious repercussions for negative outcomes.
   Frequently_____ occasionally_____ rarely_____
3. You and other managers are adept at making timely maneuvers to block rivals from taking over your market position.
   Frequently_____ occasionally_____ rarely_____
4. You and your fellow managers act with boldness, which has a positive psychological impact on employee behavior and morale.
   Frequently_____ occasionally_____ rarely_____

5. Your firm is organized to support and facilitate communication among staff for rapid action on time-sensitive market opportunities.
Frequently_____ occasionally_____ rarely_____
6. Recognizing the value of your employees' knowledge, skill, and prior experience, you actively encourage them to tap their expertise for reliable decisions.
Frequently_____ occasionally_____ rarely_____

Part 2: Symptoms of Strategy Guideline 1 Functioning Ineffectively

1. You are stalled by a cloud of complacency and apprehension within your group, which results in a lack of fresh ideas and initiatives for growing the business.
Frequently_____ occasionally_____ rarely_____
2. Your personnel are overly preoccupied with defending an existing market position. Negligible amounts of time and effort are spent searching for new market and product opportunities.
Frequently_____ occasionally_____ rarely_____
3. Other managers and staff are caught up by fear of what competitors might do. Such negative behavior discourages any effort to mount a vigorous response strategy to counter competitors' efforts.
Frequently_____ occasionally_____ rarely_____
4. You find yourself exhibiting undue caution, which permeates the group and prevents action on new market opportunities.
Frequently_____ occasionally_____ rarely_____
5. The organization tends to lose momentum.
Frequently_____ occasionally_____ rarely_____

Part 3: Remedial Actions Based on Parts 1 and 2

Parts 1 and 2 provide qualitative assessments of your ability to *shift to the offensive*. Based on your self-diagnosis or that of your team, use the following remedies to apply corrective actions.

Attack complacency: Develop ready-to-implement contingency plans that you can call upon to seize market opportunities or blunt competitive threats.
Prepare for new opportunities: Use market and competitor intelligence to support decisions and reduce the risks inherent in making bold moves.

Instill confidence: Conduct specialized training related to enhancing employees' job skills using real-world competitive scenarios, improving discipline, and elevating their self-assurance and morale.

Regain momentum: Employ market segmentation to understand how and where to deploy resources at a decisive point, thereby increasing your chances of success against a numerically stronger competitor. You thereby reduce the risks of moving into heavily defended markets.

Encourage an entrepreneurial mind-set: Require key personnel to submit ideas that would take the company into new markets, products, and services. Use the Strategy Action Plan that follows for this purpose.

Support a market-driven orientation: Build a cross-functional strategy team that encourages creativity and visionary thinking.

Prepare for unexpected events: Hold reserves.

Remedial action points: _____

_____

Reference: Chapter 8, Turn Uncertain Market Situations into Fresh Possibilities: Move to the Offensive.

Executive Comments on Guideline 1

To illustrate the practicality of Guideline 1, "Shift to the Offensive," the following actual quotes come from a survey of chief executives of medium and large organizations. (Company names are withheld to maintain confidentiality.) The same survey is used for Executive Comments in the guidelines that follow.

"We're facing an avalanche of competition and bogged down by erratic marketing."

"My company is losing momentum and losing its way."

"Need to find the way back to the market where we got creamed."

Strategy Guideline 2: Maneuver by Indirect Strategy

An indirect strategy applies strength against a competitor's weakness, resolves customer problems with offerings that outperform those of your competitors, and achieves a psychological advantage by creating an unbalancing effect in the mind of your rival manager.

Part 1: Indications That Strategy Guideline 2 Functions Effectively

1. You intentionally integrate indirect strategies into your business plans, thereby increasing the success rate of your efforts. You also engage in open dialogues with colleagues and staff about new approaches to indirect strategies, along with their implementation.
   Frequently_____ occasionally_____ rarely_____
2. You act with the understanding that acquiring the skills to implement indirect strategies opens your mind to fresh ideas. You thereby reduce the risks of going after market leaders, even where limited resources are available.
   Frequently_____ occasionally_____ rarely_____
3. You deliberately employ indirect approaches that distract the competing manager into making false moves and costly mistakes.
   Frequently_____ occasionally_____ rarely_____
4. You intentionally avoid getting entangled in direct confrontations with competitors, which would result in the unnecessary draining of resources.
   Frequently_____ occasionally_____ rarely_____

Part 2: Symptoms of Strategy Guideline 2 Functioning Ineffectively

1. You fail to develop indirect strategies that outthink, outmaneuver, and outperform competitors'.
   Frequently_____ occasionally_____ rarely_____
2. You neglect to probe for unserved market niches where there is minimal resistance from competitors—and where opportunities exist to establish a foothold and expand into a mainstream market.
   Frequently_____ occasionally_____ rarely_____
3. Personnel do not rely on competitor intelligence to formulate an indirect strategy.
   Frequently_____ occasionally_____ rarely_____
4. You lack a benchmarking system to periodically evaluate strengths, weaknesses, or best practices, which can then be used to develop indirect strategies.
   Frequently_____ occasionally_____ rarely_____
5. You do not employ an organized approach, such as through a cross-functional team, to tap the diverse backgrounds of individuals and

develop indirect approaches to enter markets or defend against an aggressive competitor.

Frequently_____ occasionally_____ rarely_____

Part 3: Remedial Actions Based on Parts 1 and 2

Parts 1 and 2 provide qualitative assessments of your ability to maneuver *by indirect strategy*. Based on your self-diagnosis or that of your team, use the following remedies to apply corrective actions:

Use a SWOT (strengths, weaknesses, opportunities, threats) examination, or other comparative analysis tool to help you determine which indirect strategies to employ.

Institute checkpoints to confirm that your indirect strategies are moving you from your current competitive situation toward the objectives you want to achieve. Make shifts quickly and assertively according to your findings.

Use all available sources of intelligence to interpret your market position. Such input provides additional clues to the development of your indirect strategies.

Find an unattended, poorly served, or emerging market segment as a decisive target in which to implement an indirect strategy for market expansion.

For ideas on how to configure a strength-against-weakness mix that cannot be easily matched by competitors, see Chapter 1, Table 1.1, "Selecting the Extraordinary and Normal Forces to Maneuver by an Indirect Strategy."

Remedial action points: _____

Executive Comments on Guideline 2

"Need to outmaneuver local and regional rivals and try to muscle into new markets."

"We must find a way to move closer to the end user and develop a true competitive edge."

"We must fend off competition not only from low-cost providers, but also from competitors with more flexible products."

Strategy Guideline 3: Act with Speed

Nothing drains resources like an overlong campaign. There are few cases of prolonged operations that have been successful. Drawn out efforts often divert interest, diminish enthusiasm, and damage employee morale.

Part 1: Indications That Strategy Guideline 3 Functions Effectively

1. You realize that dragged-out campaigns have rarely been successful. You work to avoid them in your team and organization before they divert interest, depress morale, and deplete resources.
   Frequently _____ occasionally _____ rarely _____

2. You recognize that speed is an essential component to securing a competitive lead. This impacts market share, product positioning, and ultimately customer relationships.
   Frequently _____ occasionally _____ rarely _____

3. Your staff understands that even minor delays can result in a loss of momentum and could signal a vigilant competitor to move in and fill the void.
   Frequently _____ occasionally _____ rarely _____

4. You and your staff know that a strategy that integrates speed with technology puts you in an excellent position to secure a competitive lead.
   Frequently _____ occasionally _____ rarely _____

5. You recognize that speed adds vitality to a company's operations and becomes a catalyst for growth.
   Frequently _____ occasionally _____ rarely _____

6. You understand that acting defensively to protect a market position is but a preliminary step to moving boldly and rapidly against a competitor.
   Frequently _____ occasionally _____ rarely _____

Part 2: Symptoms of Strategy Guideline 3 Functioning Ineffectively

1. You and your staff fail to understand that excessive delay acting on time-sensitive market conditions can result in losses in market share and competitive position.
   Frequently _____ occasionally _____ rarely _____

2. A general malaise exists in the organization, which results in missed opportunities.
   Frequently _____ occasionally _____ rarely _____
3. Personnel lack initiative in implementing business plans with speed.
   Frequently _____ occasionally _____ rarely _____
4. You are slow in preventing a product from reaching a commodity status, frequently resulting in price wars.
   Frequently _____ occasionally _____ rarely _____
5. Insufficiently trained and inexperienced staff has resulted in an absence of organizational vitality, thus hindering movement forward.
   Frequently _____ occasionally _____ rarely _____
6. Despite suitable market conditions, you are unable to act boldly and with speed.
   Frequently _____ occasionally _____ rarely _____
7. You have failed to secure a competitive lead due to sluggishness in integrating technology into the marketing mix.
   Frequently _____ occasionally _____ rarely _____
8. Organizational layers prolong deliberation and delay decisions, creating a trickle-down corporate culture of procrastination.
   Frequently _____ occasionally _____ rarely _____
9. There is a persistent lack of urgency in developing new products.
   Frequently _____ occasionally _____ rarely _____

Part 3: Remedial Action Based on Parts 1 and 2

Parts 1 and 2 provide qualitative assessments of your ability to *act with speed*. Based on your self-diagnosis or that of your team, use the following remedies to apply corrective actions:

Reduce organizational obstacles that prevent you and your staff from increasing the speed of internal communication and of acquiring feedback from the field.

Require selected individuals on your staff to submit timely proposals with the prime objective of creating additional revenue streams through product innovations, new technology applications, and by identifying new, unserved, or poorly served market segments.

Use a cross-functional strategy team to take maximum advantage of a diversity of backgrounds and perspectives as well as gain buy-in to business plans.

Actively seek input from managers and field personnel in pinpointing competitors' weaknesses and areas of vulnerability.

Conduct training to break down internal barriers and areas of friction that reduce speed.

Remedial action points: _____

_____

Reference: Chapter 1, Apply Strength against Weakness: Maneuver by Indirect Strategy.

Executive Comments on Guideline 3

"Products are late and customers are holding off purchases of existing products, which is slowing the company's revenue growth. The delays can only help the competition."

"Delayed models damaged morale and created a dramatic loss of market share."

"We must outmaneuver local and regional rivals by bringing sophistication more rapidly to the market than they can."

Strategy Guideline 4: Grow by Concentration

Adopt strategies that concentrate resources at a decisive point where you can gain superiority in select areas. That includes targeting a competitor's specific weakness or a general area of vulnerability.

Part 1: Indications That Strategy Guideline 4 Functions Effectively

1. You recognize that concentration is a strategy with which to challenge larger competitors using a segment-by-segment approach.
   Frequently _____ occasionally _____ rarely _____
2. You and your colleagues feel competent about concentrating your resources to gain a superior position in a selected market segment, even if it creates some exposure elsewhere.
   Frequently _____ occasionally _____ rarely _____

3. You use market intelligence to pinpoint a segment for initial market entry. You then use that position to expand toward additional growth segments.
Frequently _____ occasionally _____ rarely _____

4. You and your staff are adept at reaching beyond traditional demographic and geographic segmentation approaches, employing more advanced classifications to identify new or underserved segments.
Frequently _____ occasionally _____ rarely _____

5. You and others are flexible about extracting your organization from underperforming segments and concentrating on faster growing ones.
Frequently _____ occasionally _____ rarely _____

6. You are capable of concentrating your resources against competitors' weaknesses (the indirect approach).
Frequently _____ occasionally _____ rarely _____

7. Marketing and sales personnel are skilled at recommending products and services to suit the specific needs of their customers.
Frequently _____ occasionally _____ rarely _____

8. You understand that every market segment presents opportunities to fill market gaps, allocate resources efficiently, and exploit a rival's limitations. And your organization has a practice of seeking out and uncovering them.
Frequently _____ occasionally _____ rarely _____

Part 2: Symptoms of Strategy Guideline 4 Functioning Ineffectively

1. You fail to select market segments that offer long-term growth.
Frequently _____ occasionally _____ rarely _____

2. You dissipate resources across too many segments.
Frequently _____ occasionally _____ rarely _____

3. Your company's product launches disappoint in the absence of a strategy focused on segment needs.
Frequently _____ occasionally _____ rarely _____

4. Your people have not internalized the principle that achieving a competitive edge means employing a strategy of concentration; yet it does not expose your own pockets of vulnerability or spread resources too thinly.
Frequently _____ occasionally _____ rarely _____

Part 3: Remedial Actions Based on Parts 1 and 2

Parts 1 and 2 provide qualitative assessments of your ability to use a *strategy of concentration*. Based on your self-diagnosis or that of your team, use the following remedies to apply corrective actions:

Install an ongoing competitor intelligence system to identify a competitor's weaknesses.

Concentrate on emerging markets or those that are poorly served in order to get a foothold into additional segments.

Secure your position with dedicated services and customized products that would create barriers to competitors' entry.

Within customer segments, tailor products and services built around product differentiation, value-added services, and business solutions that exceed those of competitors.

Conduct internal strategy training sessions, especially for those individuals who do not understand the value of finding a decisive point and resist adopting a strategy of concentration.

Remedial action points: _____

Reference: Chapter 3, Secure a Competitive Advantage: Concentrate at a Decisive Point.

Executive Comments on Guideline 4

"We need to develop small, quasi-independent units, each of which focuses on its particular market with freedom to do what it takes to win."

"We must drive down costs, while beefing up R&D and finding new places to grow."

"We have to find ways to woo back angry retailers and build the brand image among groups that were once behind the product."

"Need to develop an organization to yield a new and far more competitive company."

Strategy Guideline 5: Prioritize Competitor Intelligence

Worse than no information, is wrong information. Commit to a practice of gathering and elevating competitor intelligence to the highest standard. Seek information that can impact your decisions about selecting markets, launching new products, and devising competitive advantages. Reliable intelligence helps you outthink, outmaneuver, and outperform your competitor.

Part 1: Indications That Strategy Guideline 5 Functions Effectively

1. An evaluation meeting takes place after each major competitive encounter, which includes (a) comparing competitors' performance against yours, (b) assessing strategies and tactics that worked or failed, and (c) identifying the most useful type of competitor intelligence.
Frequently _____ occasionally _____ rarely _____

2. Your company provides adequate funding for competitor intelligence and makes it a key part of the strategy development process.
Frequently _____ occasionally _____ rarely _____

3. You and others use competitor intelligence as an evaluation tool for assessing the levels of risk associated with strategy options.
Frequently _____ occasionally _____ rarely _____

4. You use some type of comparative analysis to determine strong points and areas of vulnerability compared with those of competitors.
Frequently _____ occasionally _____ rarely _____

5. You use competitor intelligence to increase accuracy in selecting markets, locating an optimum position, and determining how to defend against a competitor's intrusion.
Frequently _____ occasionally _____ rarely _____

6. You and others recognize that competitor intelligence provides a way to anticipate rivals' moves and to devise strategies to outmaneuver them, enabling you to use your resources to maximum efficiency.
Frequently _____ occasionally _____ rarely _____

Part 2: Symptoms of Strategy Guideline 5 Functioning Ineffectively

1. Competitors' actions frequently catch you by surprise, hampering your ability to respond with speed and effectiveness.
Frequently _____ occasionally _____ rarely _____

2. Lack of real-time information about market events leads to indecisiveness, which filters down, affecting the attitudes and morale of your staff.
Frequently _____ occasionally _____ rarely _____

3. Personnel are inclined to misjudge, exaggerate, or underestimate a competitor's situation due to unreliable or insufficient back-up intelligence.
Frequently _____ occasionally _____ rarely _____

4. There is a tendency to develop product launch plans without documenting market conditions, competitors' strengths and weaknesses, and buyers' specific needs.
Frequently _____ occasionally _____ rarely _____

5. Your organization lacks a flexible channel of two-way communication between the field and the home office, which would flag opportunities or threats.
Frequently _____ occasionally _____ rarely _____

Part 3: Remedial Actions Based on Parts 1 and 2

Parts 1 and 2 provide qualitative assessments of your overall ability to *prioritize competitor intelligence*. Based on a self- or team diagnosis of your company's situation, use the following remedies to implement corrective action:

Integrate competitor intelligence into your business planning process. Use it as the centerpiece for developing offensive and defensive strategies.

Train staff with various functions—in particular, the sales force—to recognize the changing dynamics of consumer needs and competitive threats as well as the urgency for them to actively gather reliable intelligence.

Use a team approach, as well as agents, to track competitors' activities, with particular attention to decoding their strategies.

Set up a procedure for competitive benchmarking to upgrade your firm's systems and processes.

Use outside sources to expand the reach of your competitor intelligence activities.

Remedial action points: _____

Reference: Chapter 4, Create a Lifeline to Business Strategy: Employ Competitor Intelligence.

Executive Comments on Guideline 5

"Sluggish markets, ferocious competition, and our management's own ineptness pummeled the organization."

"Upstart clone makers beat us with low-cost products through new channels, such as the Internet, phone, and superstores."

"We need to keep the company from becoming an also-ran in the industry. But we don't have enough quality information to guide us."

Strategy Guideline 6: Tune in to Your Corporate Culture

Corporate culture is the operating system and nerve center of your organization. It forms the backbone of your business strategy and shapes how your employees react in a variety of internal and external situations. It guides employee reaction in the face of a crisis. Understand it, tap into its strengths, and reenergize your company.

Part 1: Indications That Strategy Guideline 6 Functions Effectively

1. You actively tune in to your company's traditions, values, beliefs, and history. You acknowledge that corporate culture is at the heart of your operation and is at the core of what makes the organization tick.
   Frequently _____ occasionally _____ rarely_____

2. You realize that your organization's culture shapes how employees think, act, and are likely to perform in a variety of competitive encounters.
   Frequently _____ occasionally _____ rarely _____

3. You recognize corporate culture as the DNA that shapes all competitive strategies. It serves as a primary determinant in selecting strategies and tactics that will succeed.
   Frequently _____ occasionally _____ rarely_____

4. You are aware that corporate culture influences your leadership style and consequently your ability to implement business plans.
   Frequently _____ occasionally _____ rarely_____

5. You acknowledge that corporate culture, as one of the prime differentiators, gives your organization a unique identity among customers and competitors.
   Frequently _____ occasionally _____ rarely _____

6. You attempt to study your competitors' corporate cultures to reveal their inner workings. Thus, you can predict with some accuracy how rivals will react under a variety of market conditions.
Frequently _____ occasionally _____ rarely _____

Part 2: Symptoms of Strategy Guideline 6 Functioning Ineffectively

1. Personnel operate within a closed-in (and consequently uninspiring) culture that prevents them from recognizing the hard-nosed realities of global competition.
Frequently _____ occasionally _____ rarely _____
2. The existing culture is passive—staff fails to internalize the consequences of falling behind in new technology; similarly, they do not recognize the potential of new technology for exploiting fresh opportunities.
Frequently _____ occasionally _____ rarely _____
3. The corporate culture is incompatible with the changing dynamics of the marketplace. And your personnel do not bend with shifts in customers' buying behavior.
Frequently _____ occasionally _____ rarely _____
4. You fall short in aligning business plans and corporate culture, thereby jeopardizing the plans themselves.
Frequently _____ occasionally _____ rarely _____
5. At times you go against the core values, beliefs, and historical traditions that represent the organization's deep-rooted culture.
Frequently _____ occasionally _____ rarely _____

Part 3: Remedial Actions Based on Parts 1 and 2

Parts 1 and 2 provide qualitative assessments of your overall ability to tune in to your *corporate culture*. Based on a self-diagnosis or that of your team, regarding your company's situation, use the following remedies to implement corrective action:

Determine if the existing culture is compatible with the long-term vision and objectives of the organization.

Create change—either rapidly forced or gradually nurtured. In so doing, use your organization's symbols, signs, and rituals to shift and reinterpret your culture without abandoning your core values.

Determine which intrinsic qualities of your organization's culture are unique and could form a distinctive identity for use as a differentiation strategy.

Through a cross-functional team, a total customer orientation, and open communication, utilize the diverse talents of your employees to shape a vigorous corporate culture that is in tune with the global competitive environment.

Consciously align your business plan with your corporate culture.

Remedial action points: _____

_____

Reference: Chapter 5, Maintain High Performance: Align Competitive Strategy with Your Company's Culture.

Executive Comments on Guideline 6

"An element of arrogance in our company's culture leads some managers to ignore the intense competition."

"We need to energize management and smooth over customers."

"We must meld clashing cultures and heal the rifts."

Strategy Guideline 7: Develop Leadership Skills

The pillars of true leadership are insightfulness, straightforwardness, compassion, strictness, and boldness. Develop your own personal style of leadership that fits your character and personality, suits the job requirements, and harmonizes with the organization's culture.

Part 1: Indications That Strategy Guideline 7 Functions Effectively

1. You display superior leadership skills by openly communicating to your personnel a clear vision, purpose, and direction for the business unit or company.

   Frequently _____ occasionally _____ rarely _____

2. You understand that leadership means inspiring your people, developing strategies, organizing actions, and responding to market and competitive issues rapidly and effectively.

   Frequently _____ occasionally _____ rarely _____

3. You score high in interpersonal skills, as well as in the ability to recognize the inherent dignity and worth of those you manage.
Frequently _____ occasionally _____ rarely _____

4. As a leader, you demonstrate to your staff an expertise in and devotion to developing competitive strategies.
Frequently _____ occasionally _____ rarely _____

5. You reveal leadership in motivating your employees to win. This means winning customers and sustaining a long-term profitable position in the marketplace.
Frequently _____ occasionally _____ rarely _____

6. You show superior leadership by helping subordinates grow and succeed through ongoing training and coaching.
Frequently _____ occasionally _____ rarely _____

7. You recognize that there is no single leadership style. You foster a competitive environment with a personalized and flexible managerial style.
Frequently _____ occasionally _____ rarely _____

Part 2: Symptoms of Strategy Guideline 7 Functioning Ineffectively

1. Personnel display a hint of mistrust in your ability to assess market and competitive conditions and thereby in your ability to make timely and accurate decisions.
Frequently _____ occasionally _____ rarely _____

2. Employees exhibit negative behavior, confusion, and an unwillingness to take the initiative.
Frequently _____ occasionally _____ rarely _____

3. The organization or business unit is enveloped by a malaise that creates stress and anxiety.
Frequently _____ occasionally _____ rarely _____

4. You do not have an effective system of recognition and rewards.
Frequently _____ occasionally _____ rarely _____

5. Employees show fear of loss: loss of pride in the organization, loss of status, loss of the respect of their peers, or possibly loss of employment.
Frequently _____ occasionally _____ rarely _____

6. Subordinates are rarely asked for input, and feedback is seldom offered.
Frequently _____ occasionally _____ rarely _____

7. Your overall leadership style tends to be inconsistent with the company's core values.

    Frequently _____ occasionally _____ rarely _____

Part 3: Remedial Actions Based on Parts 1 and 2

Parts 1 and 2 provide qualitative assessments of your overall ability to use *leadership skills*. Based on a self- or team diagnosis of your company's situation, use the following remedies to implement corrective action:

Communicate to staff a vision for the firm (or business unit) so that they can picture their individual roles as relevant to the organization's long-term outlook.

Permit individuals to act on their own initiative. Create and enforce a feedback loop that serves the organization and the individuals who comprise it.

Motivate employees to improve their skills via ongoing training and individual coaching.

Create unity of effort through discipline and the fostering of a team effort.

Develop a leadership style that harmonizes with the culture of the organization, its objectives and strategies.

Remedial action points: _____

_____

Reference: Chapter 6, The Force Multiplier behind Your Business Strategy: Leadership.

Executive Comments on Guideline 7

"The business is floundering due to a lack of cohesiveness and focus."

"We need to energize the product line plagued by missed opportunities, sluggish sales, and dismal profits."

"Need to keep the company from becoming an also-ran in the industry."

Strategy Guideline 8: Create a Morale Advantage

An organization is only as strong as its people. Morale tends to energize individuals with the resolve to act decisively under competitive pressures.

As a dominant form of behavioral expression, focus on employee morale, discipline, and trust to strengthen your chances of winning in competitive strategy.

Part 1: Indications That Strategy Guideline 8 Functions Effectively

1. You actively seek to heighten morale and use it to energize staff, tap their inner strengths, and inspire them to act decisively under competitive pressures.
   Frequently _____ occasionally _____ rarely _____
2. You recognize that high morale holds a team together, affects day-to-day performance, and ultimately contributes to how well a business plan is implemented.
   Frequently _____ occasionally _____ rarely _____
3. You understand that even with cutting-edge technology and outstanding business strategies, your efforts will languish if employees are not roused with the driving will to succeed.
   Frequently _____ occasionally _____ rarely _____
4. You recognize that to be successful as a manager, you need to boost morale by the following means: (a) emphasizing ethical standards of behavior, (b) encouraging constructive communication, (c) rewarding individuals for meaningful performance, and (d) using cross-functional teams to encourage innovation.
   Frequently _____ occasionally _____ rarely _____
5. With morale affecting day-to-day employee performance, you try to create a work environment that fosters creative thinking and encourages employees to offer opinions and ideas—with the assurance they will be given serious attention.
   Frequently _____ occasionally _____ rarely _____
6. You move rapidly to overturn employee skepticism about the organization's commitment to meaningful change; and you do so before it festers into a morale problem.
   Frequently _____ occasionally _____ rarely _____

Part 2: Symptoms of Strategy Guideline 8 Functioning Ineffectively

1. Employees visibly display a lack of respect toward leadership.
   Frequently _____ occasionally _____ rarely _____

2. Employees seem to feel insecure about their jobs and have doubts about management's concern for their well-being.
   Frequently _____ occasionally _____ rarely _____
3. Some key managers do not appear to be concerned about morale and their employees are seemingly left to fend for themselves.
   Frequently _____ occasionally _____ rarely _____
4. There is no sign of a cohesive team spirit.
   Frequently _____ occasionally _____ rarely _____
5. There is a conspicuous lack of new ideas, innovations, or opportunities bubbling up to management, as well as a noticeable absence of a workable two-way communication system.
   Frequently _____ occasionally _____ rarely _____
6. No consistent procedure is present to encourage and reward employees for innovative suggestions.
   Frequently _____ occasionally _____ rarely _____
7. Employees exhibit erratic behavior; appear discouraged, indifferent, fearful; and resist risk taking.
   Frequently _____ occasionally _____ rarely _____
8. Where new market initiatives and competitive encounters have failed, management routinely lays blame primarily on poor employee performance.
   Frequently _____ occasionally _____ rarely _____

Part 3: Remedial Actions Based on Parts 1 and 2

Parts 1 and 2 provide qualitative assessments of your overall ability to *build morale*. Based on a self- or team diagnosis of your company's situation, use the following remedies to implement corrective action:

Communicate a long-term positive vision for the organization with clearly stated objectives that employees can internalize and get excited about.

Develop a learning environment that demonstrates to employees management's interest in and support for their development.

Remove personal and physical barriers that would prevent employees from feeling pride in their work and the organization.

Create collaborative cross-functional teams with specific duties and responsibilities.

Encourage constructive communication up and down the organization where self-expression and innovation are encouraged.

Nurture a corporate culture that positively impacts morale, especially when it is reinforced by an ethical climate and secured by the company's positive history and core values.

Utilize media tools such as social networking to identify and capitalize on this rapidly growing venue for marketing, messaging, and competitive edge.

Require teams to submit strategy action plans. This serves as a unifying activity to allow collaboration and encourages the flourishing of team dynamics through creative expression.

Remedies and actions: _____

_____

Part 1 reference: Chapter 7, "Engage Heart, Mind, and Spirit: Create a Morale Advantage."

Executive Comments on Guideline 8

"With consolidation throughout the industry and a more competitive environment, there is a need to squeeze savings, cut jobs, and close branches. Yet, it's imperative that we keep employee morale up and reassure customers that they won't be abandoned."

"We've stumbled badly because of shoddy quality. Now we must get our people behind the product line, play down the tarnished name, improve quality, and set prices a bit below the market leaders."

Strategy Guideline 9: Strengthen Your Decision-Making Capabilities to Think like a Strategist

Leaders use history, market events, and their intuition as a guide to sharpening decision making and strategy skills. In the end, it is mastering the concepts and principles of strategy that can direct marketplace events in your favor.

Part 1: Indications That Strategy Rule 9 Functions Effectively

1. You actively strengthen your decision-making capability by systematically examining the best practices of other companies and using them to build additional layers of knowledge.

Frequently _____ occasionally _____ rarely _____

2. Where reliable competitive intelligence is lacking, you tend to rely with confidence on your innate intuition, experience, training, and knowledge to arrive at valid decisions.
Frequently _____ occasionally _____ rarely _____

3. To enhance the decision-making skills of front-line employees, you give them ample opportunity and free reign to develop their imaginations and intuitions.
Frequently _____ occasionally _____ rarely _____

4. You understand that by valuing your organization's business history and by managing knowledge, you add precision to your decision-making skills.
Frequently _____ occasionally _____ rarely _____

Part 2: Symptoms of Strategy Guideline 9 Functioning Ineffectively

1. You lack an organized effort to capture and record the lessons and key strategies of past events. This inhibits the ability to pass them on to the next generation of decision makers.
Frequently _____ occasionally _____ rarely _____

2. You do not have a systematic procedure or venue for individuals to share experiences, insights, knowledge, and observations.
Frequently _____ occasionally _____ rarely _____

3. Intense competitive situations cause staff to flounder and become dispirited.
Frequently _____ occasionally _____ rarely _____

Part 3: Remedial Actions Based on Parts 1 and 2

Parts 1 and 2 provide qualitative assessments of your overall ability to improve the knowledge and the decision-making capabilities of individuals to think like strategists. Based on a self- or team diagnosis of your company's situation, use the following remedies to implement corrective action:

Provide training sessions where key individuals are invited to provide their insights and experiences.

Set up databases that capture case histories and details of significant business events, as well as noteworthy competitive encounters—both failed and successful.

Permit individuals to access the assembled information for their personal use, to coach others, and for formal training to think like strategists.

Remedies and actions: _____

_____

Part 1 reference: Chapter 9, "Think like Strategists: Lessons from the Masters of Strategy."

Executive Comments on Guideline 9

"Merchandise isn't moving and customers are shopping elsewhere."
  "We're zigzagging between two conflicting goals: market share or profits."
  "We need experience in joint ventures, mergers, and other types of alliances if we're going to succeed as a global competitor."

Summary

This Strategy Diagnostic Tool provides a three-dimensional view consisting of (1) the human element that elevates the natural worth of people, (2) the company's culture that determines if plans can succeed, and (3) strategies that decide the success of a product line, business unit, or company.

  If you merge them into a solid platform that characterizes you and your leadership, there will be ample opportunities for growth.

Appendix 2: Appraising Internal and External Conditions

APPRAISING INTERNAL AND EXTERNAL CONDITIONS

Used as an audit, the appraisal consists of 100 questions to provide an accurate assessment of your company's operating condition and level of competitiveness.

Taking the time to conduct the evaluation reduces some of the risks of planning, strategizing, and implementing your Strategy Action Plan. For instance, you can anticipate the weaknesses and strengths from both your side and that of your competitor and make informed decisions based on analysis and fact, not speculation.

The aim of the audit, therefore, is to highlight a set of symptoms for further evaluation. Then, with the detailed output, you can take corrective actions or modify plans to meet those circumstances. Thus, the appraisal permits you to conduct a structured analysis of internal and external considerations divided into three areas:

1. Your firm's market environment
2. Management procedures and policies
3. Strategy factors

Where possible, use a team approach in conducting the audit to gain varied perspectives on key issues. Also, if feasible, you will find it beneficial to add objective outside opinions from individuals when evaluating your organization's competencies. Should you encounter some questions that do not apply to your business, unless they can be modified to fit your purpose, move on to the next section.

Appraising Internal and External Conditions

Part 1: Reviewing the Firm's Market Environment

Consumers (End Users)

1. Who are our ultimate buyers?
2. What are the primary physical features and psychological influences in their buying decisions?
3. What are the key factors that make up the demographic and psychographic (behavioral) profiles of our buyers?
4. When and where do they shop for and consume our product? With what frequency are purchases made?
5. What needs do our products or services satisfy?
6. How well do they satisfy compared with similar offerings from competitors?
7. How can we segment our target markets effectively? (See advanced techniques for selecting a segment's decisive point in Chapter 3.)
8. How do prospective buyers perceive our product? Is there valid research to support the opinions?
9. What are the economic conditions and expectations of our target market over the near and far term?
10. Are our consumers' attitudes, values, or habits changing? Is there a reliable gauge to track changes?

Customers (Intermediaries)

11. Who are our intermediate buyers, such as distributors or retailers? How influential are they?
12. How well do those intermediaries serve our target market?
13. How well do we serve their needs? What are the special areas that would help solidify relationships?
14. What are the central issues that drive their buying decisions? Have there been any changes over the past 12 months? What are the projections for the next 12 months?
15. Has there been any significant movement in their locations relative to the end users?
16. What types of noncompeting products do they carry? What competing product lines do they handle?
17. What percentage of total revenue does each competing product line represent compared with ours?
18. How much support do they give our product? What can be improved?

19. What factors made us select them and them select us?
20. How can we motivate them to work harder for us?
21. Do we need them?
22. Do they need us?
23. Do we use multiple channels, including e-commerce?
24. Would we be better off setting up our own distribution system?
25. Should we go direct? What are the advantages and disadvantages?

Competitors

26. Who are our competitors? What are their strengths and weaknesses?
27. Where are they located relative to our key customers?
28. How big are they overall and, specifically, in our product areas? Where are they vulnerable?
29. What is their product mix? Are there any gaps in their mix that would create an opportunity for us?
30. Is their participation in the market growing or declining? What are the reasons in either situation?
31. Which competitors may be leaving the field? Why? What are the implications for us?
32. What new domestic competitors may be entering the market? What market niches are they filling? Or what comparative advantage are they using?
33. What new foreign competitors may be on the horizon? How threatening are they?
34. Which competitive strategies and tactics appear particularly successful or unsuccessful when used by competitors—and by us?
35. What new directions, if any, are competitors pursuing? Give details of new strategies as they relate to markets, products, pricing, marketing, supply-chain, technologies, leadership, or other.

Other Relevant Environmental Components

36. What legal and environmental constraints affect our marketing efforts? What are the immediate and long-term concerns?
37. To what extent do government regulations restrict our flexibility in making market-related decisions?
38. What do we have to do to comply with regulations?
39. What political or legal developments are looming that will improve or worsen our situation?

40. What threats or opportunities do advances in technology hold for our company?
41. How well do we keep up with technology in day-to-day operations? How do we rank in the industry and against key competitors?
42. What broad cultural shifts are occurring in segments of our market that may impact our business?
43. What consequences will demographic and geographic shifts have for our business?
44. Are any changes in resource availability foreseeable, for example, finances, equipment, personnel, raw materials, suppliers?
45. How do we propose to cope with environmental, social, or "green" issues that can impact our business?

Part 2: Reviewing Management Procedures and Policies

Analysis

46. Do we have an established market research function or rely on outside resources? To what extent are agents used to obtain intelligence?
47. Do we systematically use competitor intelligence for developing plans and strategies? Is it applicable in determining an indirect approach, pinpointing strengths and weaknesses of competitors, and focusing on a decisive point at which to commit resources?
48. Do we subscribe to any regular market data services?
49. Before we introduce a new product or service, do we test its acceptance among customers, as well as consider how it is positioned against competitors?
50. Are all our major market-related decisions based on market research and competitor intelligence?

Planning

51. Do we have a formalized procedure to develop a Strategic Business Plan or Strategy Action Plan?
52. To what extent do we seek collaboration in planning by opening the process to additional levels of the organization and thereby obtain qualified feedback that leads to higher levels of morale and results in employee buy-in to implement the plan?
53. Do our long- and short-term objectives complement our company's mission or overall strategic direction?

54. What market-driven procedures do we use to locate gaps in customers' needs?
55. Do we develop clearly stated short-term and long-term objectives? How are they prioritized so as to avoid internal conflicts among other business units?
56. Are our objectives realistic, achievable, and measurable?
57. Do we utilize appropriate metrics to assess performance and make essential mid-course corrections?
58. How effective are we in integrating what-if scenarios into our plan and thereby are prepared to react rapidly to competitive challenges and threats to our business?
59. Are our core strategies and tactics for achieving our objectives aligned with our corporate culture?
60. Is there a systematic screening process in place to identify opportunities and threats, such as a SWOT analysis or similar approach?
61. How aggressively are we considering diversification or joint ventures as these relate to planning for growth?
62. How effectively are we segmenting our target market to determine decisive points?
63. Are we committing sufficient resources to accomplish our objectives?
64. Are our resources optimally allocated to the major elements of our marketing mix?
65. How well do we tie in our plans with the other functional plans of our organization?
66. Is our plan realistically followed or just filed away?
67. Do we continuously monitor environmental movements to determine the adequacy of our plan?
68. Do we have a centralized activity to collect and disseminate market and competitor intelligence?
69. Do we have an individual who oversees the sharing of technology and marketing data among business groups?

Organization

70. Does our firm have a high-level function to analyze, plan, and oversee the implementation of our strategic efforts?
71. How capable and dedicated are our personnel?
72. Is there a need for more internal skills and leadership training?
73. Are our managerial responsibilities structured to best support the needs of different products, target markets, and sales territories?

74. Does our organization's culture embrace and practice the market-driven, customer-first concept?

Part 3: Reviewing Strategy Factors

Product Policy

75. What is the makeup of our product mix?
76. How effective are our new product development plans?
77. Does it have optimal breadth and depth to maintain customer loyalty and prevent unwelcome entry by aggressive competitors?
78. Should any of our products be phased out?
79. Do we carefully evaluate any negative ripple effects on the remaining product mix before we make a decision to phase out a product?
80. Have we considered modification, repositioning, and/or extension of sagging products?
81. What immediate additions, if any, should be made to our product mix to maintain a competitive edge?
82. Which products are we best equipped to make ourselves, and which items should we outsource and resell under our own name?
83. Do we routinely check product safety and product liability?
84. Do we have a formalized and tested product recall procedure? Is it effective?
85. Is any recall imminent?

Pricing

86. How effective are our pricing strategies? To what extent are our prices based on cost, demand, market, and/or competitive considerations?
87. How would our customers likely react to higher prices? What losses can we expect, if any?
88. Do we use temporary price promotions and, if so, how effective are they?
89. Do we suggest resale prices?
90. How do our wholesale or retail margins and discounts compare with those of the competition?

Promotion

91. Do we state our promotion objectives clearly?
92. Do we spend enough, too much, or too little on promotion?
93. Are our advertising themes and copy effective?

94. Is our media mix adjusted to the optimal use of the Internet and social media?
95. Do we make aggressive use of sales promotion techniques to stimulate sales and to disrupt competitive actions?

Personal Selling and Distribution

96. Is our sales force (if any) at the right strength to accomplish our objectives, and to what extent will the Internet affect sales force activities?
97. Is the sales force optimally organized to provide market coverage and to meet customers' logistical, technical, and service needs?
98. Is it adequately trained and motivated, and characterized by high morale, ability, and effectiveness?
99. Have we enhanced our supply-chain, or are there opportunities for further streamlining?
100. How does our overall marketing mix, including field personnel and senior management, rank against our major competitors?

As a final note: Conditions within your company are likely to change. That is unquestionably true of today's volatile markets. Therefore, to make clear-cut decisions (or recommend changes to the next level of management), it is indispensable to skillful management to give your operation a once-a-year (or sooner) checkup.

You will find the above appraisal and the Strategy Diagnostic System highly useful to assist in clarifying your thinking and permitting you to grasp the core meaning of your firm in its operating environment.

Once again, you can customize the items and questions in the checklists to conform to your company's and market's needs.

Appendix 3: Strategy Action Plan

STRATEGY ACTION PLAN

The Strategy Action Plan (SAP) is the "housing" for all your objectives, market intelligence, and ultimately the business-building strategies that will set your plan in motion.

With the forms and guidelines that follow, you can develop your personalized plan while retaining the planning structure of a proven format. You have the flexibility to customize the forms by inserting the specific terminology, vocabulary, and unique issues related to your company and industry.

You can also add your own proprietary or commercially available spreadsheet programs, as well as any of the growing number of Customer Relationship Management (CRM) programs.

As you develop your SAP, keep in mind the strategy guidelines and the lessons from the *masters of strategy* presented in this book. Use them as a continuing benchmark to strengthen your plan. They will increase the likelihood of creating new opportunities and neutralizing the effects of competitors' actions.

Planning tips: First, to the extent that you are able, make the planning process a collaborative effort. Minimally, use your own dedicated team. Where possible, involve personnel from various functions and levels of the organization. You thereby profit from a diverse group of individuals with varied knowledge, training, and experience. As a result of their involvement, morale tends to improve and there is a greater chance of gaining their enthusiastic buy-in to implementing the plan. Second, if you lack specific information to answer a question, either skip the question or use your best intuitive response. In no case, give up; keep the process going by moving on to the next item.

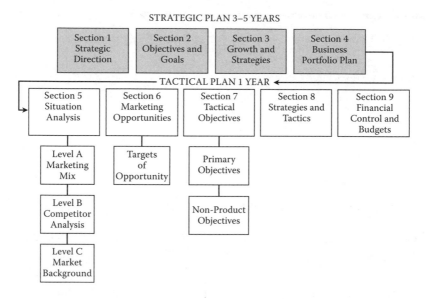

FIGURE A3.1

Overview of the Strategy Action Plan: Strategic Section

You can obtain optimum results for your SAP by following the process diagrammed in Figure A3.1. As you examine the flowchart, notice that the top row of boxes represents the *strategic* portion of the plan and covers a three- to five-year time frame.

The second row of boxes displays the *tactical* one-year plan. It is the merging of the *strategic* plan and *tactical* plan into one unified SAP that makes it a complete format and an operational management tool to energize your company's potential.

The SAP process will add an organized and disciplined approach to your thinking. Yet, in no way does it confine your thinking or creativity. Instead, the process expands your flexibility, extends your strategy vision, and elevates the innovative process. In turn, the resulting strategies provide you with a choice of revenue-building opportunities expressed through markets, products, and services.

Section 1: Strategic Direction

The first box in Figure A3.1, Section 1, *Strategic Direction,* allows you to visualize the long-term direction of your company, business unit, product line, or service.

Planning Guidelines

The following six questions provide an organized approach for you to develop a strategic direction. Your answers will help shape a vision of what your company, business unit, or product/service will look like over the next three to five years.

1. What are your firm's distinctive areas of expertise? This question refers to your organization's special competencies. The following list will trigger your thinking and help you develop an answer:

    Product or service strengths based on customer satisfaction, profitability, and market share

    Strengths as they relate to competitors

    Depth of relationships within the supply-chain and/or with end-use customers

    Existing production capabilities

    Technology advantages

    Size of your sales force

    Financial strength

    Research and development, and other product innovation initiatives

    Amount of customer or technical services provided

    Unique skills of personnel

    A corporate culture favorable to supporting the SAP

    Fill in:

    _____

    _____

2. What business should your firm be in over the next three to five years? In what ways do you see it being different from what your company or business unit looks like today?

    _____

    _____

3. What new segments or categories of customers will you serve?

    _____

    _____

4. What additional functions, products, or services will you likely offer customers as you see the market evolve?

    _____

    _____

5. What new technologies will you require to satisfy future customer and market needs, as well as maintain a favorable competitive position?

_____

_____

6. What changes will likely take place in any of the following areas that will impact your company: evolving markets and existing market segments, consumer behavior and buying patterns, availability of resources, domestic and off-shore competition, the Internet, environmental (green) issues, trade practices, and the economy? Add any other changes unique to your company or industry that will likely affect your business.

_____

_____

Now compress your answers to the above six questions into one statement that would represent a realistic strategic direction for your company, business unit, or product line.

As you develop your strategic direction, recognize that your corporate culture is the operating system and nerve center of your organization. Therefore, make certain that your company's culture can support your vision.

The following statement is an actual example of a well-written strategic direction document for a company in the medical field. Note how it casts an expansive vision for growth in a changing marketplace:

"Our strategic direction is to meet the needs of consumers and health-care providers for drug-delivery devices by offering a full line of hypodermic products and product systems. Our leadership position will be maintained through internal research and development, licensing of technology, and/or acquisitions to provide alternative administration and monitoring systems."

The company's previous statement indicated a narrower and highly restrictive vision, which left it vulnerable to low-cost producers and with little room for opportunistic thinking and growth:

"Our position is to be a leader in the manufacture of hypodermic needles."

_____

_____

Section 2: Objectives and Goals

Planning Guidelines

State your objectives and goals both quantitatively and qualitatively (the second box in the top row of Figure A3.1).

Your primary guideline: Still using a time frame of three to five years, look again at your strategic direction, and then develop objectives that will have a wide-ranging impact on the growth of your business.

Quantitative Objectives

Indicate major performance expectations, such as sales growth, market share, return on investment, profit, and any other quantitative objectives required by your company.

With this longer time frame, your objectives are generally broad and relate to the total business or to a few major market segments. (In the tactical section, these objectives will be more specific for each product and market.)

_____

_____

Qualitative Objectives

Think of these objectives as a means to build on your organization's existing strengths or core competencies, or to eliminate any internal weaknesses. Above all, keep your objectives specific, actionable, realistic, and focused on developing opportunities and achieving a sustainable competitive advantage. Use the following examples to trigger objectives for your business:

Improve supply-chain relationships and/or enhance the product's position on the supply-chain.

Integrate the Internet and social networks into marketing plans.

Expand secondary distribution; for example, find innovative approaches to reach customers beyond your existing supply-chain.

Build specialty products to penetrate markets for greater market share or to discourage competitors' entry.

Establish or improve competitor intelligence procedures.

Focus training actions on improving skills, discipline, morale, and performance of employees.

Launch new products and reposition old products.

Upgrade customer relationships.

Expand technical services.

Improve marketing mix (product, price, promotion, and supply-chain) management.

_____

_____

Section 3: Growth Strategies

Planning Guidelines

This section outlines the process you use to achieve your objectives and goals. Think of *strategies* as actions to achieve your longer-term objectives; *tactics* as actions to achieve shorter-term objectives.

Because this time frame covers three to five years, strategies are indicated here. The one-year portion, illustrated later in the plan, identifies tactics.

In practice, where you have developed broad-based, long-term objectives, you should list multiple strategies for achieving each objective. In instances where you find it difficult to apply specific strategies, it is appropriate to use general strategy statements.

Suggestions: How you write your strategies can vary according to your individual or team's style. For example, you can restate each objective from Section 2, followed by a listing of corresponding strategies.

Still another option is to write a general strategy statement followed by a detailed listing of specific objectives and corresponding strategies.

The key point: Each objective (what you want to accomplish) must be followed by one or more strategies (actions to reach your objective.)

Additionally, as you develop your strategies, keep in mind the guidelines presented in this book: *indirect strategy, speed, concentration at the decisive point, bold action, competitor intelligence, alignment of strategy with corporate culture,* and *leadership.*

You can also use the following suggested categories in which to apply the strategy guidelines:

Markets at different stages of the life cycle—introduction, growth, maturity, or decline

Products at different stages of the life cycle—introduction, growth, maturity, or decline

Brand development

Competitive positioning

Comparative advantage in such areas as quality, value-added options, services, technology, or guarantees

Market share expansion

Market share defense against incoming competitors

Supply-chain options

Internet and social media

Financing and asset allocation

Specific marketing, technology, and manufacturing strengths to be exploited

Employee training and development

Corporate culture, people management, and internal operating systems

---

Section 4: Business Portfolio Plan

Planning Guidelines

The business portfolio includes listings of *existing* products and markets, as well as *new* products and markets. Following a logical progression, it is based on the strategic direction, objectives and goals, and growth strategies outlined in the previous sections.

In particular, your portfolio should mirror your strategic direction. That is, the broader the scope of your strategic direction, the more expansive the range of products and markets in the portfolio.

Conversely, the narrower the dimension of your strategic direction, the more limited the content of products and markets.

Use the following format to develop your own business portfolio:

Existing products/existing markets (market penetration):
    List those *existing* products you currently offer to *existing* customers or market segments. In an appendix of the SAP, you can document sales, profits, market share data, and other pertinent facts related to growth potential or competitive issues. You can then determine if your level of penetration is adequate and if possibilities exist for further growth.

New products/existing markets (product development):

    In this section, extend your thinking and list potential *new* products or services you can offer to *existing* markets. Again, recall the guideline that the broader the dimension of your strategic direction, the broader the possibilities for the content of your portfolio. Also, continue thinking in a time frame of three to five years.

_____

_____

Existing products/new markets (market development):

    Now list *new* markets for your existing products. Explore possibilities for market development by identifying emerging, neglected, or poorly served segments in which existing products can be utilized.

_____

_____

New product/new markets (diversification):

    This portion of the business portfolio is somewhat visionary, because it involves developing *new* products for *new* and yet untapped markets. Consider new technologies, global markets, the green trend, and potential alliances. Once again, interpret your strategic direction in its broadest context. Do not seek diversification for its own sake. Rather, the whole purpose of the exercise is for you to develop an organized framework for meaningful expansion.

_____

_____

After identifying any new opportunities, it may be necessary for you to revisit Section 3 (Growth Strategies) and list actions you would take to implement the opportunities.

The Business Portfolio completes the strategic portion of your SAP. Now you are ready to proceed to the one-year tactical plan.

The Strategic Business Plan

Tactical Section: Overview

The tactical plan, the second row of boxes in Figure A3.2, designated as Sections 5 through 9, is not a stand-alone plan. It is an integral part of the total SAP.

FIGURE A3.2

Where commonalties exist among similar products and markets, one tactical plan can work as long as you make the appropriate changes in such areas as the sales force and the communications mix.

Where you face substantial differences in the character of your products and markets, develop separate tactical plans.

Suggestion: Avoid the temptation to develop your plan by jumping into the middle of the SAP and beginning the planning process with the one-year tactical plan. There are no suitable shortcuts, because primary input to the tactical plan is a logical flow of reflective thinking and calculated information from the strategic portion of the SAP (top row) of boxes. It all forms the foundation for the situation analysis, opportunities, annual objectives, tactics, and budgets that follow in the second row of boxes.

Section 5: Situation Analysis

The following three-part situation analysis details the past and current situations of your business:

Level A: Marketing mix (product, pricing, supply-chain, and communications)
Level B: Competitor analysis
Level C: Market background

The purpose of the situation analysis is to define your business in a factual and objective manner. Compile historical data for a period of at least three years.

Doing so provides an excellent perspective about where your company has been, where it is now, and where you want it to go as defined in your Strategic Direction (Section 1).

Level A: Marketing Mix—Product

Planning guidelines: Objectively describe the performance of your product or service by: sales history, profitability, share of market, and any other required metrics. Where appropriate, chart sales history with spreadsheets, graphics, or your company's forms.

_____

_____

Explain the product's current position as it relates to market penetration, reputation, and competitiveness. Also, indicate where the product is in its life cycle, for example, introduction, growth, maturity, or decline.

_____

_____

Identify future trends as they relate to such areas as the environment, technology, industry, customers, and competitive factors that may affect future prospects.

_____

_____

Indicate the intended purpose of your product in terms of its applications or uniqueness.

_____

_____

Describe the features and benefits of your product as they relate to quality, performance, reliability, safety, convenience, or any other factors important to customers. Use the same factors as comparisons with competitive offerings.

_____

_____

List other pertinent product information, such as expected product improvements and additional product characteristics, for example, size, model, price, or packaging; recent features that enhance the product's position; trends in features, benefits, technological changes; and changes that would add superior value to the product and provide a competitive advantage.

_____

_____

Level A: Marketing Mix—Pricing

History of pricing: Examine the history of pricing strategies for each market segment and describe its impact on the product's market position.

_____

_____

Future pricing trends: Indicate pricing trends as they pertain to new product features, expected market changes, trade and consumer reactions, and competitive responses to your pricing strategies.

_____

_____

Level A: Marketing Mix—Supply-Chain

Current channels: Describe the makeup of your supply-chain. Identify the functions performed at each stage within the network (distributor, dealer, direct, e-commerce). Indicate levels of performance, expressed in sales volume, profitability, and market share.

Where appropriate, also analyze your physical distribution system, such as warehouse locations, inventory systems, transportation, or just-in-time delivery procedures.

_____

_____

Effectiveness of coverage: Characterize the effectiveness of coverage by the programs and services provided to each channel. Comment on the overall effectiveness of the supply-chain. Specify the key activities performed at each point and indicate any areas that require corrective action. Include comments on current practices in supply-chain management, including e-commerce.

_____

_____

Future trends: Indicate future trends in supply-chain management and how they would affect various links in the chain. Also, describe trends in e-commerce and the need for new methods of physical distribution.

_____

_____

Level A: Marketing Mix—Communications: Advertising,
Sales Promotion, Internet, Social Media, Publicity, Sales

Planning guidelines: Evaluate your efforts directed at each market segment or distribution channel based on the following: expenditures, creative strategy, social media, types of promotion, Internet, and other forms of communications unique to your industry.

_____

_____

Sales force functions: List functions performed by your company's (or distributor's) sales force. Indicate their effect on various market segments. If applicable, comment on such approaches as a "push" strategy (through distributors) or a "pull" strategy (through end users.)

_____

_____

Key accounts: List key accounts and their levels of performance. Comment on their contributions to long-term growth, sales, and profits. Add any comments related to providing special services and resources to retain them.

_____

_____

Strategies: Identify your company's past and current communication strategies by product and market segment, and describe trends in those areas.

_____

_____

Other support strategies: Identify other support programs (publicity, educational, professional, trade shows, webinars, literature, films/videos, and social media) that you have used and evaluate their effectiveness.

_____

_____

Competitive trends: Identify and evaluate competitive trends in the same categories as above. Look at off-shore as well as domestic competitors.

_____

_____

Level B: Competitor Analysis

Planning guidelines: Make your competitive analysis as comprehensive as possible. The more competitive intelligence you gather and use, the more strategy options you have open to you.

Market share: List all your competitors in descending order by size, along with their sales and market shares, if applicable.

Include your company's ranking within the listing. Show at least three competitors, more if the information is meaningful. (If you are unable to provide usable information for this portion, you should give high priority to developing a competitor intelligence effort.)

_____

_____

Competitors' strengths and weaknesses: Identify each competitor's strengths and weaknesses related to such factors as product development, distribution, pricing, promotion, management leadership, caliber of employees, and financial condition. (Refer to Chapter 1, Table 1.1 for a more comprehensive list of criteria from which to develop an indirect strategy.)

Indicate any significant trends that would signal unsettling market situations, such as aggressive moves by a competitor to grow market share to maintain its market position.

_____

_____

Product competitiveness: Identify competitive pricing strategies, including discount practices, if any. Identify those competitors firmly entrenched in low-price segments of the market, those at the high end of the market, or competitors that are lodged in dedicated niches.

_____

_____

Product features and benefits: Compare the specific features and benefits of your product against those of competing products. In particular, focus on product quality, design factors, and performance. Evaluate price/value relationships for each, discuss customer preferences, and identify unique product innovations.

_____

_____

Advertising effectiveness: Identify competitive spending levels and their effectiveness, as measured by awareness levels, competitive copy claims, and reach/frequency levels (if available.)

Such measurements are often conducted by advertising agencies, marketing research firms, or some publications. If you cannot use these sources, rely on informal observation and whatever measurements you can obtain.

_____

_____

Effectiveness of the supply-chain: Compare competitive distribution strengths and weaknesses. Address differences in market penetration, market coverage, delivery time, and physical movement of the product by regions or territories.

_____

_____

Packaging: Compare packaging based on performance, innovation, and customer preference. Also review size, shape, function, convenience of handling, ease of storage, and shipping.

_____

_____

Trade/consumer attitudes: Review both trade and consumer attitudes toward product quality, warranties, customer/technical service, company image, and company performance.

_____

_____

Competitive share and market trends: Specify trends in market share by individual products, as well as by market segments. Further, identify where each competitor is making a major commitment and where it may be relinquishing control by product and segment.

_____

_____

Sales force effectiveness and market coverage: Compare effectiveness by sales, service, frequency of contact, and problem-solving capabilities by competitor and by market segment against your own performance. Also, look at all sales force performance within the supply-chain.

Level C: Market Background

Planning guidelines: This last part of the situation analysis focuses on the demographic and behavioral factors of your market. Here is where you determine market size and customer preferences—both business-to-business and business-to-consumer.

If you give careful attention to compiling accurate information, you can provide useful input to the following sections of the SAP: Section 6, Opportunities; Section 7, Objectives; and Section 8, Strategies/Tactics. This information also highlights any gaps in knowledge about markets and customers, which helps you determine what additional market intelligence is needed to make more effective decisions.

Fill in the following categories as part of the market background.

Customer profile: Describe existing and potential customers that you or your intermediaries serve. Profile them by type of products used, level of service required, price sensitivity, and promotional support. Also look at customer purchases by frequency, volume, and seasonality of purchase.

Additional information might include issues related to maintaining customer inventory levels, just-in-time delivery, retail-stocking policies, and volume discounts. Also consider e-commerce and any resulting impact on buying behavior.

Geographic aspects of products used: Define customer purchases by region and any limitations with regard to territory you encounter.

Market characteristics: Assess the demographic, psychographic (lifestyle), and other relevant characteristics of your customers.

Look at buyers' purchase patterns and any distinctive types of behavior, as well as attitudes toward your company's products, services, quality, and image.

_____

_____

Decision maker: Define who makes the buying decisions, and when and where decisions are made. Identify the various individuals or departments that may influence the decision.

_____

_____

Customer motivations: Identify the key motivations that drive your customers to buy your product. Why do they select your company over a competitor?

Consider such factors as quality, performance, price, company image, technical/customer service, convenience, location, delivery, access to upper-level management, friendship, or peer pressure.

_____

_____

Customer awareness: Define the level of consumer awareness of your products. To what extent do they:

Recognize a need for your product?
Identify your product, brand, or company as a possible supplier?
Associate your product, brand, or company with desirable features?

_____

_____

Segment trends: Define trends in the size and character of the various segments or niches you serve. A segment should be considered if it is accessible, measurable, potentially profitable, and has long-term growth potential.

Segmenting a market also serves as an offensive strategy to identify emerging, neglected, or poorly served markets that can catapult you to further sales growth. You can also consider segments as part of a defensive strategy to prevent a potential competitor from making inroads in your market through an unattended market niche at a decisive point.

_____

_____

Other comments or critical issues: Add general comments or references that would contribute to your knowledge of the market and customer base. Also list any critical issues that have surfaced as a result of conducting the situation analysis—ones that should be singled out for special attention.

Section 6: Market Opportunities

Planning Guidelines

In this section, examine strengths, weaknesses, and options. Opportunities will begin to emerge as you consider a variety of alternatives.

Try to avoid restricted thinking. Take your time and brainstorm. Dig for opportunities with other members of your planning team. Ideally, if the team includes individuals from different functional areas of the business, you will gain a diverse range of viewpoints.

Consider all possibilities for expanding existing market coverage and laying the groundwork for entering new markets. Also consider opportunities related to outthinking and outmaneuvering your competition with indirect strategies.

For instance, offensively, which competitors can you displace from which market segments? Defensively, which competitors can you deny entry into your market space?

As you go through this section, revisit your strategic portion of the SAP (top row of boxes in Figure A3.1). Also refer to the situation analysis in Section 5, specifically the competitive analysis, which will reveal voids or weaknesses that could turn into opportunities.

Use the following screening process to identify your major opportunities and challenges. Once you identify and prioritize the opportunities, convert them into objectives and tactics, which form the action topics for the next two sections of the SAP.

Present markets: Identify the best opportunities for expanding present markets by

Cultivating an additional revenue stream through new users
Displacing the competition
Increasing product usage or services by present customers
Redefining market segments
Reformulating or repackaging the product

Identifying new applications for the product

Repositioning the product to create a more favorable perception by consumers, which can turn into a competitive advantage over rival products

Expanding into emerging, neglected, or poorly served market niches

_____

_____

Targets of opportunity: List any areas outside your current market segment or product line, not included in the above categories, that you would like to explore. Be innovative and entrepreneurial in your thinking.

These areas are opportunistic. Therefore, due to their bold and possibly higher-risk characteristics, they are isolated from the other opportunities. Those you select for special attention are placed in a separate part of the objectives section of the SAP.

_____

_____

Section 7: Tactical Objectives

Now list the objectives you want to achieve during the current planning cycle— generally defined as a 12-month period to correspond with annual budgeting procedures.

Once again, you will find it useful to review Sections 5 and 6. Also, it will be helpful to review the strategic portion of the plan in Sections 1 through 4.

You want to be certain that the actions you select relate to your long-range strategic direction, objectives, strategies, and business portfolio— and those are incorporated into your tactical one-year objectives.

Tactical objectives consist of three parts:

Assumptions: projections about future conditions and trends

Primary objectives: metrics related to what you want to accomplish, including targets of opportunity

Functional objectives: operational and functional goals representing various parts of the business

Planning Guidelines: Assumptions

For objectives to be realistic and achievable, you must first generate assumptions and projections about future conditions and trends. List only those major assumptions that will affect your business for the planning period.

Economic assumptions: Comment on the overall economy, local market economies, consumer expenditures, and changes in customer buying patterns. Also document any impact on market size, growth/decline rates, costs, and trends in your market segments.

_____

_____

Technological assumptions: Include the likelihood of technological breakthroughs and applications of new technology to your business.

_____

_____

Sociopolitical assumptions: Indicate any positive and restrictive legislation, political tensions, tax outlook, population patterns, and educational factors. Also list changes in customer habits linked to social media, green issues, and e-commerce.

_____

_____

Competitive assumptions: Identify activities of existing competitors, inroads of new competitors, with particular attention to surging global competitors, and subsequent changes in trade practices.

_____

_____

Planning Guidelines: Primary Objectives

Focus on the primary financial objectives that your organization requires. Also include targets of opportunity that you initially identified as innovative and entrepreneurial in Section 6.

Where there are multiple objectives, you may find it helpful to rank them in order of priority. Be sure to use metrics to quantify expected results where possible. You can separate your objectives into the following categories:

Primary objectives: Current and projected sales, profits, market share, return on investment, and other quantitative measures. (Use either Table A3.1, forms provided by your organization, or any commercial spreadsheet software.)

Targets of opportunity objectives:

_____

_____

TABLE A3.1

Primary Objectives

Product Group Breakdown	Current				Projected			
	Sales ($)	Units	Margins	Share of Market	Sales ($)	Units	Margins	Share of Market
Product A								
Product B								
Product C								
Product D								

Planning Guidelines: Functional Objectives

State the *functional objectives* relating to both product and nonproduct issues in each of the following categories. Should any of the categories not apply to your current business situation, consider applications for the future, and then move on.

Product objectives:

Quality: List quality objectives that could achieve a competitive advantage.

_____

_____

Development: Set objectives to obtain new technology by exploring internal research and development, licensing, or joint ventures.

_____

_____

Modification: Describe major or minor product changes through reformulation, redesign, or reengineering.

_____

_____

Differentiation: Identify objectives to differentiate the product by delivering new applications to reach new customer groups within existing markets, or by expanding into additional geographic areas.

_____

_____

Diversification: Indicate technology transfer objectives to drive new product development.

_____

_____

Deletion: List products to be removed from the line due to unsatisfactory performance, and those to be kept in the line for strategic reasons, such as presenting your company to the market as a full-line supplier.

_____

_____

Segmentation: List potential line extensions (adding product varieties) to reach new market niches, or to defend against an incoming competitor in an existing market segment.

_____

_____

Pricing objectives:
Show list prices, volume discounts, and promotional rebates.

_____

_____

Promotion objectives:
Indicate sales force support, sales promotion, webinars, advertising, Internet, social media, and publicity to the trade and consumers.

_____

_____

Supply-chain objectives:
Identify potential new intermediaries to increase geographic coverage and to solidify relationships with the trade; list distributors or dealers to be removed from the chain.

_____

_____

Physical distribution objectives:
Identify logistical factors, from order entry to the physical movement of a product through the supply-chain, and eventual delivery to the end user.

_____

_____

Packaging objectives:
List functional design and/or decorative considerations for brand identification.

_____

_____

Service objectives:
Identify a broader range of services, from providing customers access to key personnel in your firm, to providing on-site technical assistance.

_____

_____

Other objectives:
Indicate other objectives as suggested in Targets of Opportunities.

_____

_____

Planning Guidelines

Nonproduct objectives:
Although most activities eventually relate to the product or service, some are support functions that you should consider. (Obtain input from personnel in other functions, as needed.)

Key accounts:
Indicate those customers with whom you can develop special relationships through customized products, distribution, value-added services, or participation in quality improvement programs.

_____

_____

Manufacturing:
Identify special activities that would provide a competitive advantage, such as offering small production runs to accommodate the changing needs of customers and reduce inventory levels.

_____

_____

Marketing research:
Cite any customer studies or market data that reveal opportunities for new revenue streams.

_____

_____

Credit:
Include any programs that use credit and finance as a value-added service, such as providing financial advice or offering financial assistance to customers in specific situations.

_____

_____

Technical sales activities:
Include any support activities, such as 24/7 hot-line assistance or on-site consultation to solve customers' problems.

_____

_____

Research and development:
Indicate internal research and development projects, as well as joint ventures that would complement the Strategic Direction identified in Section 1 of the SAP.

_____

_____

Human resource development and training:
Identify specialized training and development programs to upgrade the skills of those individuals who are responsible for implementing the SAP.

_____

_____

Other:
Include any activities that would contribute to your organization's uniqueness, and thereby provide an indirect strategy to achieve a competitive advantage.

_____

_____

Section 8: Strategies and Tactics

Overview

Strategy is the art of coordinating the means (money, human resources, materials) to achieve the ends (profits, customer satisfaction, growth) as defined by company policy, strategic direction, and objectives.

From another perspective, strategies are actions to achieve long-term objectives; tactics are actions to achieve short-term objectives.

Therefore, in this section strategies and tactics are identified and put into action. Responsibilities are assigned, schedules set, budgets established, and checkpoints determined.

Make sure that individuals involved with developing and implementing the plans actively participate in this section.

Planning Guidelines

Restate the functional product and nonproduct objectives from Section 7 and link them to the strategies and tactics you will use to reach each objective.

One of the reasons for restating the objectives is to clarify the frequent misunderstanding between objectives and strategies. Objectives are *what* you want to accomplish; strategies are actions that specify *how* you intend to achieve your objectives.

Note: If you state an objective and do not have a related strategy, you may not have an objective. Instead, the statement may be an action for some other objective.

Objective 1: _____

    Strategy/Tactic _____

Objective 2: _____

    Strategy/Tactic _____

Objective 3: _____

    Strategy/Tactic _____

Planning Guidelines

Summary strategy: Summarize the basic strategies for achieving your primary objectives. Include alternative and contingency plans should you come up against unexpected situations that prevent you from reaching your objectives.

Note: You may choose to repeat this information as an *Executive Summary* in the beginning of the plan.

As you develop your final strategy statement, use the following checklist to determine its completeness. Also add those that are specific to your industry and company.

Changes to the product or package, including differentiation and value-added services

Strategies that create a competitive advantage, along with contingency plans to block competitors' aggressive moves

Changes to price, discounts, or long-term contracts that impact your market share

Changes to marketing strategy, such as the selection of features and benefits, or copy themes to special groups

Strategies to reach new, poorly served, or unserved market segments—along with indications of decisive points for entry and defense

Promotion strategies aimed at dealer and/or distributor, consumer, and sales force

Internal changes in the operating systems, as well as initiatives that would better align the corporate culture with your strategies

_____

_____

Section 9: Financial Controls and Budgets

Planning Guidelines

Having completed the strategy phase of your SAP, you must decide how you will monitor its execution. Therefore, before implementing it, develop procedures for both control (comparing actual and planned figures) and review (deciding whether planned figures should be adjusted or other corrective measures taken).

This final section incorporates your operating budget. If your organization has reporting procedures, you should incorporate them within this section.

Included below are examples of additional reports or data sheets for you to consider. They are designed to monitor progress at key checkpoints of the plan and to permit either major shifts in strategies or simple midcourse corrections:

Forecast models related to the industry, environment, competition, and any other areas that are applicable to your company and your plan

Sales by channel of distribution, including
  Inventory or out-of-stock reports
  Average selling price (including discounts, rebates, or allowances), along the supply-chain and by customer outlet

Profit-and-loss statements by product

Direct product budgets

Research and development budget

Administrative budget

Spending by quarter

As an overall guideline—regardless of the forms you use—make certain that the system serves as a reliable feedback mechanism.

Your interest is in maintaining explicit and timely control so you can react swiftly to impending problems. Further, it should serve as a procedure for reviewing schedules and strategies.

The only other section in your SAP is an appendix. It should include the following items:

Relevant industry data and market research that provide information on trends, product usage, market share, and the like

Data on competitors' strategies, including supporting information on their products, pricing, promotion, distribution, market position, as well as profiles of management leadership (if available)

Details about your new product features and benefits

In addition, various computer databases, as well as a wide variety of customized or off-the-shelf software programs, are available to assist in monitoring and strengthening your plan.

Final Guidelines

After completing your plan, take a thoughtful look at its key parts. For instance, review the breadth of your strategic direction and be sure it impacts the objectives, strategies, and portfolio sections that follow.

As for strategies, check to see that you have utilized the *Lessons from the Masters of Strategy* cited in this book, such as developing an indirect approach, acting with speed, concentrating on a decisive point, and moving to the offensive. Also, be sure you have backed up your strategies with solid competitor intelligence.

At the same time, be absolutely certain that you have aligned the plan with the culture of your organization. That means you have taken into consideration your company's history, its behavioral patterns, and leadership; the attitudes, temperaments, and skills of employees; their capacity to confront aggressive competitors; and their ability to live with risk. Also, determine if you have sufficient resources, including financial, technology, and physical, to implement the plan.

Finally, to implement the plan takes leadership and your ability to reach the hearts and minds of the individuals who are going to make it all happen. Good luck!

Index